BASIC WIRING

Other Publications:

AMERICAN COUNTRY
VOYAGE THROUGH THE UNIVERSE
THE THIRD REICH
THE TIME-LIFE GARDENER'S GUIDE
MYSTERIES OF THE UNKNOWN
TIME FRAME
FIX IT YOURSELF
FITNESS, HEALTH & NUTRITION
SUCCESSFUL PARENTING
HEALTHY HOME COOKING
UNDERSTANDING COMPUTERS
LIBRARY OF NATIONS
THE ENCHANTED WORLD
THE KODAK LIBRARY OF CREATIVE PHOTOGRAPHY
GREAT MEALS IN MINUTES
THE CIVIL WAR
PLANET EARTH
COLLECTOR'S LIBRARY OF THE CIVIL WAR
THE EPIC OF FLIGHT
THE GOOD COOK
WORLD WAR II
THE OLD WEST

For information on and a full description of any of the
Time-Life Books series listed above, please call
1-800-621-7026 or write:
Reader Information
Time-Life Customer Service
P.O. Box C-32068
Richmond, Virginia 23261-2068

This volume is part of a series offering homeowners
detailed instructions on repairs, construction
and improvements they can undertake themselves.

HOME REPAIR
AND IMPROVEMENT

BASIC WIRING

BY THE EDITORS OF
TIME-LIFE BOOKS

TIME-LIFE BOOKS,
ALEXANDRIA, VIRGINIA

Time-Life Books Inc.
is a wholly owned subsidiary of
TIME INCORPORATED

Founder   Henry R. Luce 1898-1967
Editor-in-Chief   Jason McManus
Chairman and Chief Executive Officer   J. Richard Munro
President and Chief Operating Officer   N. J. Nicholas Jr.
Editorial Director   Richard B. Stolley
Executive Vice President, Books   Kelso F. Sutton
Vice President, Books   Paul V. McLaughlin

TIME-LIFE BOOKS INC.

Editor   George Constable
Executive Editor   Ellen Phillips
Director of Design   Louis Klein
Director of Editorial Resources   Phyllis K. Wise
Editorial Board   Russell B. Adams Jr., Dale M. Brown, Roberta Conlan, Thomas H. Flaherty, Lee Hassig, Donia Ann Steele, Rosalind Stubenberg
Director of Photography and Research   John Conrad Weiser
Assistant Director of Editorial Resources   Elise Ritter Gibson

President   Christopher T. Linen
Chief Operating Officer   John M. Fahey Jr.
Senior Vice Presidents   Robert M. DeSena, James L. Mercer, Paul R. Stewart
Vice Presidents   Stephen L. Bair, Ralph J. Cuomo, Neal Goff, Stephen L. Goldstein, Juanita T. James, Carol Kaplan, Susan J. Maruyama, Robert H. Smith, Joseph J. Ward
Director of Production Services   Robert J. Passantino
Supervisor of Quality Control   James King

HOME REPAIR AND IMPROVEMENT

Editorial Staff for Basic Wiring
Editor   William Frankel
Picture Editors   Adrian G. Allen, Kaye Neil Noble
Designer   Herbert H. Quarmby
Associate Designer   Robert McKee
Text Editors   Marion Buhagiar, Robert L. Tschirky
Staff Writers   Sally French, Simone Gossner, Lee Hassig, Kumait Jawdat, Michael Luftman
Researchers   Brian McGinn, Ginger Seippel, Scot Terrell
Copy Coordinators   Ricki Tarlow, Eleanor van Bellingham
Art Associates   Faye H. Eng, Kaye Sherry Hirsh, Richard Salcer
Picture Coordinator   Barbara S. Simon
Editorial Assistant   Eleanor G. Kask

Editorial Operations
Copy Chief   Diane Ullius
Production   Celia Beattie
Library   Louise D. Forstall

Correspondents: Elisabeth Kraemer-Singh (Bonn); Christina Lieberman (New York); Maria Vincenza Aloisi (Paris); Ann Natanson (Rome). Valuable assistance was also provided by: Lesley Coleman, Karin B. Pearce (London); Carolyn T. Chubet (New York); Mimi Murphy (Rome).

THE CONSULTANTS: Guy Alland, an architect and the founder of the Know-How Workshop in New York City, teaches courses in home repair and improvement. He is the co-author of *Know-How*, a book on home repair.

Harris Mitchell, special consultant for Canada, has worked in the field of home repair and improvement since 1960. He is Homes editor of *Today* magazine, writes a syndicated newspaper column, "You Wanted to Know," and is the author of a number of books on home improvement.

Louis Potts, a practical master of carpentry and electrical work, was engaged in construction projects for more than 35 years.

Ralph Scannapieco, a licensed master electrician and owner of an electrical contracting firm, was engaged in this occupation for several decades.

Bernard Shapiro is the manager of a large electrical-supply store that provides wiring equipment for both professionals and homeowners.

Stanley H. Smith, the general consultant for this book, is Associate Professor of Electrical Engineering at Stevens Institute of Technology, Hoboken, New Jersey. He has been an adviser on electrical subjects to major private and public institutions for many years.

Mark M. Steele, consultant for the revised edition of this volume, is a professional home inspector in the Washington, D.C., area and an editor of home improvement articles and books.

Library of Congress Cataloging in Publication Data
Basic wiring / by the editors of Time-Life Books.—Rev. ed.
   p. cm.—(Home repair and improvement)
   Includes index.
ISBN 0-8094-7362-3
ISBN 0-8094-7363-1 (lib. bdg.)
1. Electric wiring, Interior—Amateur's manuals.
I. Time-Life Books   II. Series.
TK9901.B38   1989
621.319'24—dc19   89-4381 CIP

Contents

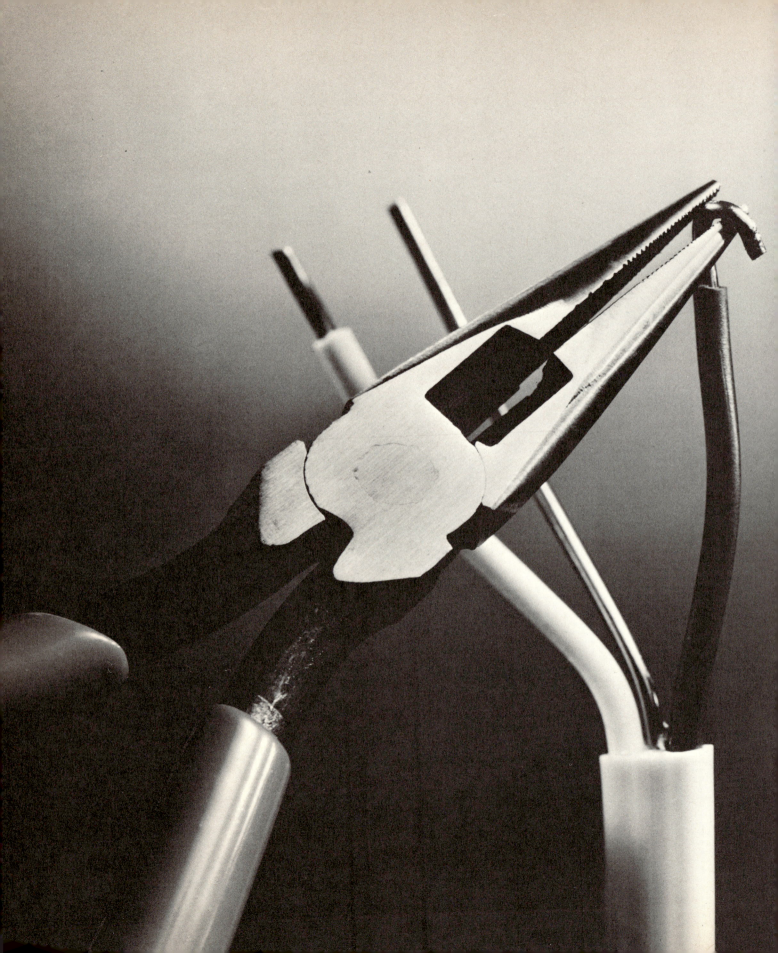

A Guide for the Home Electrician

Preparing an installation. Long-nose pliers bend a conductor wire for attachment to a screw terminal in the most common operation in electrical work. The plastic-sheathed cable, which contains three copper wires, is the kind that is used in most house wiring.

Working with wiring is something that many homeowners are unwilling to do for themselves. Electricity appears to be mysterious, even baffling; electrical jobs seem to hold an aura of danger. Like many other misconceptions, these beliefs are built up from grains of truth—or from no truth at all.

Take the idea that electrical jobs are complex and laborious. In reality, they are straightforward, orderly and, in the great majority of cases, surprisingly easy. Procedures and equipment are standardized. The techniques of wiring actually call for less manual skill than most other home repair jobs. When you are replacing a defective switch in a wall, for instance, the job involves a few turns of a screwdriver, a stop at a hardware or electrical supply store, and about 10 minutes of work—all at a cost of a dollar or two.

Some homeowners have a vague feeling that working on their own wiring is illegal—and some municipalities do insist upon permits for certain kinds of wiring. But not one of the replacement and repair jobs in this book requires a permit. If you install new wiring by extending an existing circuit or adding a new one, you may need a permit, and your work may have to be approved by an electrical inspector. Check with your municipality's building department about local regulations concerning the specific job you plan to do.

Another common misconception is that working on your home's electrical wiring will automatically endanger your fire-insurance coverage. The typical insurance policy has no provision to prohibit—or permit—electrical work you do yourself. However, if a fire were caused by wiring you installed, the company might decide that you were a poor risk and prove reluctant to renew your policy later on.

Probably the biggest deterrent of all to do-it-yourself wiring is fear —the fear of causing a shock or fire. Here the grain of truth in the common belief is real. Electricity can indeed be dangerous—but only when it is improperly handled. The danger is controlled if you follow the safety rules that are listed on page 23.

More than any other system in your home, your electrical system is standardized. The types and sizes of electrical equipment, and the methods by which they are installed, are established in the United States by the National Electrical Code and in Canada by the Canadian Electrical Code. Local codes, which have the force of law, generally follow the national codes. The watchdogs of electrical equipment are Underwriters' Laboratories, Inc. and, in Canada, the Canadian Standards Association, both independent, nonprofit organizations. The tests these groups perform are for the sole purpose of making sure that electrical devices are free of defects that can cause fire or shock. When you buy supplies, look for the UL or CSA label to be certain you get equipment that has been tested.

Boxes, wires, switches and so on come in standard sizes and shapes so that one manufacturer's products are interchangeable with another's; by matching the data printed on a device, you can be sure of buying a replacement that will fit. Wiring methods are also standardized. Whether your home is in Calgary or Atlanta, the wires will be joined to each other and to terminals in the same way.

Though you can work on your home wiring knowing nothing about electrical theory, a casual acquaintance with some electrical terms and concepts is helpful. To explain an electrical system, an analogy—only an analogy, and not entirely accurate—is often made to a water supply system. The generator of a power company is likened to the pumping station at a reservoir, the transmission wires to the water supply mains. The electricity generator raises the voltage (that is, the electrical potential, or pressure) above ground potential, which is zero, just as the pumping station raises water to a storage tank or reservoir above ground level. In the water system gravity creates the pressure that pushes the water through the supply mains and into home piping. In the same way, voltage that is created by the generator pushes electric current through transmission lines and into the wires of house circuits.

Inside a house, electricity is supplied through "hot" wires, comparable to the pressurized supply pipes of a plumbing system. At various points along the wires are outlets in the form of lights, switches and receptacles. Turning on a light switch is somewhat like opening a faucet to let water run—an electric current flows through the hot wire to make the light glow. Once the electricity has done its work, its potential drops to zero, just as water loses pressure after flowing through a sink or laundry tub. And the electrical system has "drains" —the white neutral wires that return the current to its source—just as a plumbing system has drain pipes through which water runs into the sewer mains or the ground. In addition, the electrical system has an extra set of drains, called ground wires (page 18), that provide a fail-safe mechanism in case something goes wrong.

The light or appliance powered by the current, technically called the load, can be compared to a water wheel that remains motionless until a stream of water causes it to turn. A load may be one of two kinds. The first consists of a resistance—a material that permits the passage of electric current, but only with difficulty, and thereby creates heat. The tungsten filament of an incandescent bulb is a resistance; so is the heating element of an electric range or a coffee pot. Or a load may be an inductance—typically, a motor with windings of copper wire, in which the magnetic fields generated by the current create motion. At any moment, the demand on an electrical system depends on the number of loads in operation and their consumption of energy, just as the demand on a water system depends on how many faucets are opened and how wide they are opened.

What flows through the hot and neutral wires is, of course, not water but charged particles of matter called electrons. The wires, or conductors, offer little opposition to the movement of the particles. To prevent the escape of current, the conductors are covered with plastic or rubber insulation, blocking the passage of electricity.

The mechanics and physical fittings of the system are simple. Current moves throughout the house in wires of different sizes, according to the current a circuit may have to carry. The outlets are housed in metal or plastic boxes. All connections between wires are made inside the boxes, with plastic wire caps that twist on and off. There are boxes for receptacles, for permanently mounted light fixtures and for switches, and there are also junction boxes from which wires run in several directions. The terminals on the devices mounted in boxes are color-coded, so that you never have to guess what color of wire to attach to each terminal.

Working on any of these components is equally simple. Remember, though, to start your work by observing these two electrical safety rules: Always turn off the power before working on your wiring. Before touching any wire, test *(page 21)* to make absolutely sure you have turned off the power.

Volts, Amperes, Watts

When working on your home wiring or buying electrical devices, you will constantly come across references to the following units used to measure electricity. Each unit has a specific significance, and the first three bear a relationship to each other:

□ VOLT is the unit that measures the potential difference in electrical force, or "pressure," between two points on a circuit. The current at most receptacles and lights is at a pressure of 120 volts, although it may vary from 114 to 126. As the current moves from the hot supply wire through the load presented by an appliance or light, it loses voltage in doing work. When the current leaves the load and enters the return circuit provided by the neutral wire, it has lost all voltage and is at zero pressure, the same as the earth.

□ AMPERE is the unit used to measure the amount of current—that is, the number of electrically charged particles called electrons—that flows past a given point on a circuit each second. If you could see the particles moving along a wire and could count to 6.28 billion billion in one second, then you would have counted enough particles to make 1 ampere. Current that has lost its voltage still has amperage as it completes the circuit and returns to the power plant.

□ WATT is the unit of power. It indicates the rate at which a device converts electric current to another form of energy, either heat or motion, or to put it another way, the rate at which the device consumes energy.

□ KILOWATT-HOUR is the unit of energy, measuring the total amount of electricity that is consumed.

The relationship of volts, amperes and watts to one another is expressed in a simple equation that enables you to make any calculations you may need for projects shown in this book: Volts × amperes = watts. If the current is at 120 volts and a device requires 4 amperes of current, the equation will read: 120 volts × 4 amperes = 480 watts.

To figure the current needed for a device rated in watts, turn the equation around: Watts ÷ volts = amperes. For example, if you have an appliance, such as a toaster that uses 1200 watts: 1200 watts ÷ 120 = 10 amperes.

A Tool Kit for Basic Wiring

Tools for the electrical repair and improvement jobs in this book are few in number and require only a modest investment; some of them may already be in your home tool kit. Buy only those extra tools necessary for the specific job you are about to undertake. In addition, you will need the general-purpose tools common to most households, such as an electric drill, screwdrivers, saws, a hammer and a nail set.

☐ TESTERS. Essential to any kind of wiring job are two inexpensive testers. The voltage tester has a neon bulb and two insulated wires that end in metal probes. Its main purpose is to check that the current is off before you begin a job. It is also used, with the power on, to test for proper grounding and, in some special circumstances, to check that voltage is available in wires. Standard voltage testers can be used on circuits carrying from about 90 to 500 volts. There is also a low-voltage tester that looks similar, so be sure to check the rating on the package before you purchase one. The continuity tester has its own source of power, a small battery, and is used only with the power off to pinpoint malfunctions in a wiring component—a broken switch or light socket, for instance.

☐ MULTIPURPOSE TOOL. Preparing wires to be attached to electrical devices is accomplished efficiently with a multipurpose tool that both cuts the wires and strips insulation from them. Wire gauges printed on the tool indicate which hole to use for stripping insulation without damage to the wire. Other uses include cutting small bolts and crimping special kinds of wire connectors.

☐ PLIERS. Two kinds are needed for electrical work. Long-nose pliers have striated jaws that hold wire firmly while you shape it for attachment to a terminal. A useful model is about 7 inches long and has a wire cutter near the pivot. Lineman's pliers are used for pulling wire, bending heavier wire and twisting out removable parts of certain electrical components. These pliers also have a wire cutter and striated jaws.

☐ FUSE PULLER. If your service panel contains cartridge fuses (page 16), removing them for testing or replacement is done simply and safely with a fuse puller, which looks somewhat like pliers but, for insulating purposes, is made entirely of plastic. Fuse pullers come in one size for cartridge fuses up to 60 amperes and another for larger fuses.

☐ METAL SHEARS. Metal shears, or aviation snips, do a fast, neat job of cutting cable, and may also be needed to trim off metal flanges from certain components. Straight-cutting blades are preferable to curving blades.

☐ FISH TAPE. Two fish tapes are necessary for pulling new electrical wiring through walls. Long fish tapes are wound around a metal reel, as shown; short ones come without the reel.

☐ DRILL BITS. A ¾-inch spade bit drills a hole in wood through which cable can be run. Other sizes of spade bits may be needed for specific projects. To drill holes for running cable through masonry, use a ½-inch carbide-tipped masonry bit, the largest masonry bit that will fit a ⅜-inch drill. An 18-inch extension attachment is essential if you have to drill through thick or widely spaced beams.

☐ TAPE AND WIRE CAPS. The fasteners of electrical wiring, wire caps and electrician's tape are used for making and securing wire connections. Buy an assortment of wire caps, sized according to the gauge of the wire you are working with. If you are extending or adding a circuit, you will also need staples to fasten cable to wood studs and joists. Buy the appropriate staple for the kind and size of cable to be used.

For extending wiring outdoors, you will need a conduit bender (page 115) and perhaps a star drill (page 117), which is actually a special kind of chisel. Use it with a ball-peen hammer to make round holes for conduit in masonry. For chipping box-sized holes in masonry, use a forged-steel cold chisel.

If you plan a large wiring project, such as wiring an addition to your house, you may want to consider a couple of special tools that make the work go faster. One is an automatic wire stripper, which with one motion cuts and strips insulation. Another is a cable ripper that makes a lengthwise slit in plastic sheathing without slicing the wires and their insulation. The sheathing is then folded back and cut off with scissors or a utility knife.

VOLTAGE TESTER

CONTINUITY TESTER

WIRE CAPS

ELECTRICIAN'S TAPE

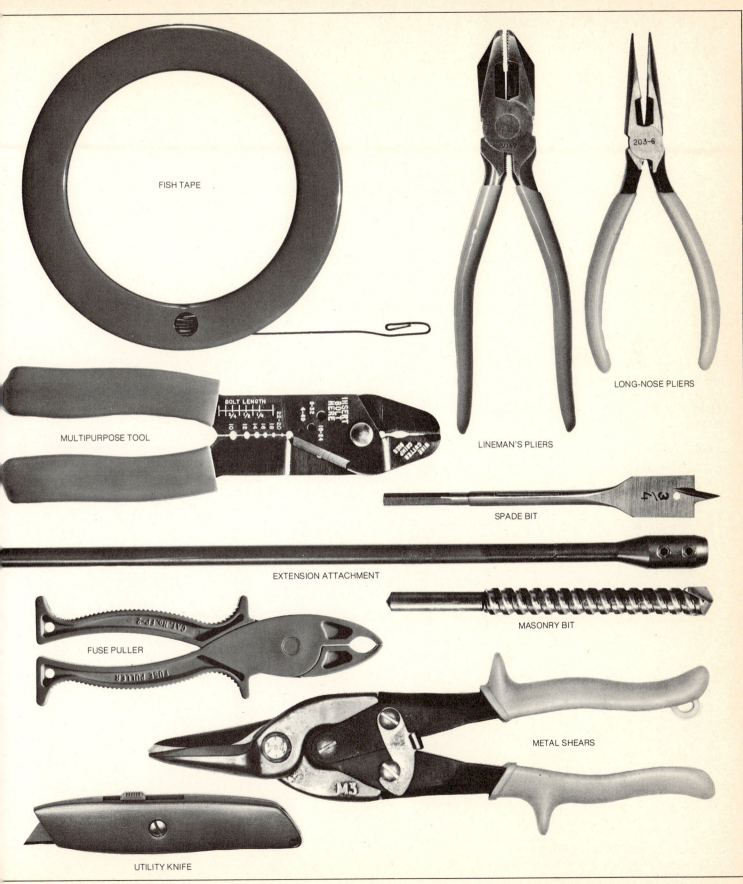

FISH TAPE

LONG-NOSE PLIERS

MULTIPURPOSE TOOL

BOLT LENGTH

INSERT BOLT HERE

LINEMAN'S PLIERS

SPADE BIT

EXTENSION ATTACHMENT

MASONRY BIT

FUSE PULLER

METAL SHEARS

UTILITY KNIFE

The Multiple Pathways of Electric Current

The wires that transmit electricity from the power company generator to your living room enter the house at a point called the service entrance. If the wires are buried beneath the ground, they enter through a pipe called a conduit. More commonly, however, they run from a power company pole on the street to a service head located high on one side of the house, as shown below. At the service head, house wires connect to the utility wires and lead down the side of the house to a meter *(page 14)* and then into the house to a service panel *(pages 15-16)*, from which power is distributed through the house by wiring systems called branch circuits.

In most homes built since World War II, the wires leading from the service head to the meter and then into the service panel are three in number. Two are hot wires that carry 120 volts each, the voltage used by branch circuits that operate lights and all but the largest appliances. For other branch circuits, the pair of 120-volt supplies is combined to make the 240 volts necessary for such large appliances as ranges and driers. The third wire is the neutral wire, maintained at zero voltage; it is connected at the service panel to a ground wire clamped to a metal rod driven into the ground or to a buried metal water pipe.

Many old homes have only two wires —one carrying 120 volts, the other neutral —coming into the service panel and thus lack 240-volt service.

In either a 120-volt or 240-volt supply system, the circuits branching from the service panel consist of cables that generally contain three or four wires enclosed in metal armor or plastic sheathing. The colors of their insulation (or the lack of insulation) denote functions.

One wire is usually bare. It serves as a safety ground, providing a direct connection between metal equipment attached to the circuit and the ground terminal at the service panel—it is meant to guarantee that a fuse will blow or a circuit breaker will open if there is a short circuit *(page 18)*.

One of the other wires is insulated in white; it is the neutral wire, at zero voltage, that completes the path of current (to make certain it can always complete the circuit, it must never be interrupted by switches, fuses or other devices).

The remaining wires are hot, each at 120 volts. If only one hot wire is in a cable, its insulation will be black; other hot wires may be red or blue, or they may be white recoded black with paint or tape at the ends. In some switch installations the white wire is hot, but it too should be recoded black *(page 43)*.

Even though the neutral wire and the ground wire are both grounded—and therefore should not give a shock—never touch them or any other wire without first testing to be certain that the wire is indeed at zero voltage *(page 21)*.

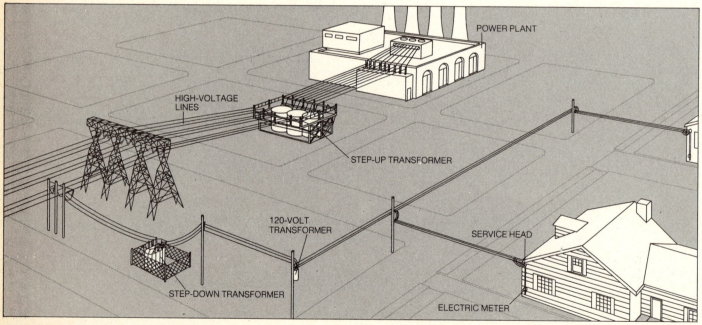

From power plant to home. On its way to your home, electricity generated at the power plant first flows through transformers that step up the voltage to as much as 765,000 volts—less energy is lost when it is transmitted at high voltage. Lines then carry the current cross-country. At points along the way, lines branch off and feed the power to step-down transformers that lower the voltage for local distribution. A final transformer lowers the voltage to 120 volts.

A variety of voltages. A heavy-duty 240-volt circuit usually serves a single large appliance like a drier or water heater. Power goes from the service panel through the cable's two hot wires, one of which is black and the other red, or sometimes white that has been recoded as hot. Each carries 120 volts to an outlet and then to the appliance. To complete the 240-volt circuit, the two hot wires alternate as the return path for each other. A bare copper wire grounds the circuit—and the appliance—by connecting the outlet box to a water pipe that goes into the earth. The range adds a fourth wire to its cable. It has the three wires —black, red and bare copper—of a 240-volt circuit for its heating elements. But it also has a

120-volt circuit for a clock and a receptacle, and it requires a white wire to provide the neutral return. The remainder of the house is served by 120-volt circuits. Most house branch circuits serve 8 to 10 outlets each, often in adjoining rooms. Such a circuit starts its "run" as a single cable. At a typical "middle-of-the-run" outlet, like that next to the TV, the incoming cable's three wires provide supply and return paths for the TV while another set of three in the outgoing cable continues the circuit to the next outlet. Other outlets, like the one for the switch near the bedroom door, may have three or more cables, each containing supply and return paths, that branch off in different directions. An "end-of-the-run" outlet

—where the shaver is plugged in—has only a single cable for the supply and return of current. The two ceiling fixtures and their switches are connected to the circuit in two ways. The bathroom light is wired like a middle-of-the-run outlet, with incoming and outgoing cables. Its third cable is a "loop" that carries current only to and from the switch. Loop wiring can be recognized by the white wire attached to one switch terminal —the wire, recoded black with tape or paint, is hot. In the bedroom, power cables enter and leave the switch outlet box. The switch terminals have only black wires connected to them— one from the incoming cable, one from the cable to the light—and no white wire is hot.

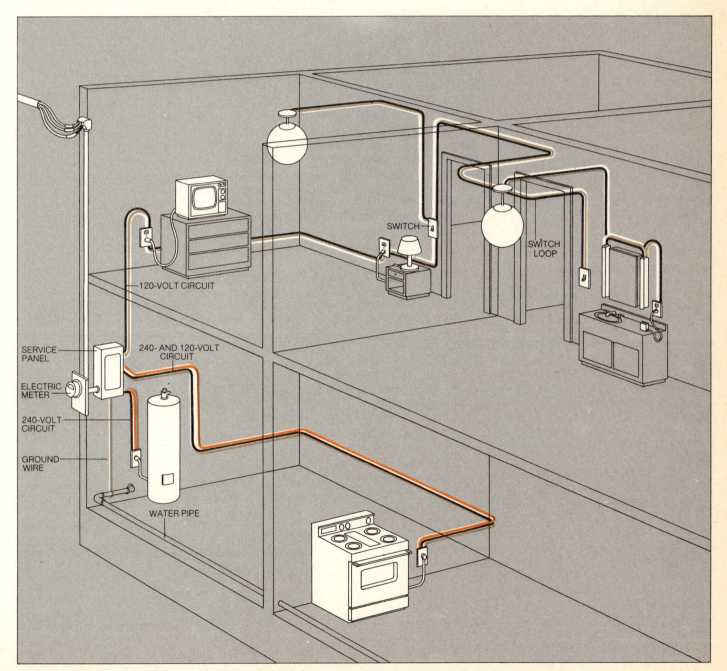

SWITCH

SWITCH LOOP

120-VOLT CIRCUIT

SERVICE PANEL

240- AND 120-VOLT CIRCUIT

ELECTRIC METER

240-VOLT CIRCUIT

GROUND WIRE

WATER PIPE

The Meter

By means of a small motor that runs faster or slower depending on how much current is entering the home, the meter registers the amount of electricity that is consumed. As the motor runs, it turns numbered indicators to register the cumulative total energy used to the time of a reading. The indicator numbers show kilowatt-hours, usually abbreviated kwh on your electricity bill. One kwh is equivalent to 1,000 watts consumed during the course of one hour.

Some meters have a row of numbers that resemble the mileage indicator in an automobile, and they are read the same way. More common, however, is the clock-type meter *(top right)* with four or five separate dials, each of which supplies one digit for the reading. The left dial registers tens of thousands, the next thousands, and so on to the right dial, which registers units, that is, single kilowatt-hours. The numbers on the dial faces are arranged alternately clockwise and counterclockwise, and the pointer moves accordingly, but no matter which direction the pointer follows, it always goes from 0 to 9.

When taking a meter reading, start with the left dial and write down the number the pointer has just passed. Be careful not to confuse the direction of the pointer. The only other difficulty in reading a meter is the ambiguity created when a pointer seems to be directly over a number—in the two sets of dials in the detail at right, the fourth dial might be read as indicating 5, but it could also indicate 4. In such a case, the clue to a correct reading is provided by looking at the adjoining dial, as explained at right.

To check your electricity bills, as you should periodically, make two consecutive readings and subtract the earlier one from the later to calculate consumption for the period between readings. Take each reading on the day the meterman's visit is due; most utility companies print this date on the bill. Then compare your results with the figures on the bill.

KILOWATT HOURS

Rr 13%

CL 200-240 V 3W •FM 2S TA 30   kh 7.2

Reading the meter. Start with the left-hand dial and write down the number that the pointer on each dial has just passed, making sure to note in each case which way the pointer rotates. This meter has a reading of 23619 kwh.

If the pointer appears to point directly to a digit, as in the fourth dial below, look at the next dial to the right. In the upper example shown, the pointer on the fifth dial has just passed 0 and clearly indicates 0. Therefore, the fourth dial should be read as indicating 5 and the two together as 50. In the lower example the pointer on the fifth dial has not yet reached 0, and it indicates 9; thus the preceding dial should be read as 4 and the two together should be read as 49. The reason is simple: the pointer of the right-hand dial must make one complete revolution before the dial to its left moves ahead one digit.

Cutting Down on Electricity Bills

You can trim ever-increasing electricity bills by installing some of the energy-saving devices described in this book:

☐ Fluorescent lights *(page 36)* provide more light than incandescent bulbs and consume less energy.

☐ Dimmer switches *(page 49)* reduce power consumption by allowing decreased lighting levels that conform to varying needs.

☐ Switches with pilot lights *(page 45)* warn when an out-of-the-way light or appliance has been forgotten.

☐ A timer *(page 45)* turns off your air conditioner when you leave home and turns it back on before you arrive.

Other ways of reducing the electricity costs do not require new equipment but simply involve changes in habits of buying or using lights and appliances:

☐ Use a single bulb of higher wattage rather than several low-watt bulbs. Two 60-watt bulbs produce less light than one 100-watt bulb though they consume about 20 per cent more energy.

☐ Use long-lasting bulbs only when the extended life is an advantage outweighing the fact that they produce less light per watt than standard bulbs.

☐ Buy lampshades that are wide at both ends: they spread light better than shades with narrow openings.

☐ Turn off incandescent lights when not in use. Although frequent switching shortens a bulb's life somewhat, it is a false notion that switching, in itself, uses power. However, you do not necessarily save by switching fluorescents off for a short time *(page 38)*.

☐ Before you buy a large appliance, see how much energy it consumes; this information usually appears on an appliance's data plate, expressed in watts. Compare this wattage with that of appliances made by other manufacturers. If all other factors are equal, choose the appliance with the lowest wattage rating. If the energy consumed is given in amperes, convert to watts by referring to the equation on page 9.

Safeguards that Shield against Dangers

In some houses, wires lead directly from the meter to a circuit-breaker panel (below); in others, the wires go to a fuse panel (page 16). At the panel the power is transferred by the main circuit breakers or fuses to the branch circuits. Each of these circuits also contains an individual breaker or fuse. Either device serves an essential safety function. It automatically interrupts the flow of current if the current exceeds the amount the circuit is designed to handle. In this way it prevents the generation of heat that could cause fire, and it also prevents shock if a short circuit connects external metal parts of the system that are grounded.

Circuit breakers are more convenient than fuses, which have to be replaced after excess current causes a circuit interruption. A breaker is quickly reset after the cause of failure has been corrected—just press the switch to "on" or in some cases to "off," then "on."

The excess current that causes a fuse to blow or a breaker to trip can arise from an overload: too many appliances or lights operating on a circuit, or an appliance drawing more power than normal. The excess current opens a switch in a circuit breaker or, in a fuse, melts a metal strip. In a ground-fault interrupter (GFI), sometimes used in place of a standard circuit breaker for circuits serving damp locations, excess current activates a microprocessor that interrupts the current, and power goes off in the entire circuit.

The same result follows a short circuit. When a bare hot wire—in the circuit itself or in an appliance, light fixture, switch or receptacle—touches a second bare wire or any metal that is grounded, current bypasses the normal circuit and flows directly to ground without having to do any work. This lack of resistance to flow makes the current very high, blowing the fuse or opening the breaker.

If a circuit breaker trips or a fuse blows, see page 20 for a way to track down the problem by calculating the load on a circuit. If your home has fuses, do not try to solve an overload problem by using a fuse with an amperage rating higher than the one that blew—it can cause dangerous overheating of the wires. Try transferring some devices to another circuit. If the problem persists, you should consider upgrading your electrical system.

Besides protecting circuits, circuit breakers and fuses perform another all-important safety function. Because fuses can be removed and breakers switched off, you can turn off the power in a circuit and work without danger of shock.

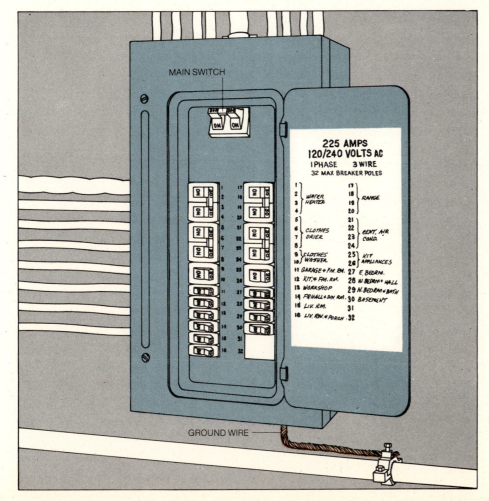

MAIN SWITCH

225 AMPS
120/240 VOLTS AC
1 PHASE   3 WIRE
32 MAX BREAKER POLES

1	17
2 WATER	18
3 HEATER	19 RANGE
4	20
	21
5	22
6 CLOTHES	23 CENT. AIR COND.
7 DRIER	24
8	
9 CLOTHES	25 KIT.
10 WASHER	26 APPLIANCES
11 GARAGE+FM.RM.	27 E.BEDRM.
12 KIT.+FM.RM.	28 N.BEDRM.+HALL
13 WORKSHOP	29 N.BEDRM.+BATH
14 FR.HALL+DIN.RM.	30 BASEMENT
15 LIV.RM.	31
16 LIV.RM.+PORCH	32

GROUND WIRE

Circuit-breaker panel. In the panel shown, a 200-ampere breaker at the top serves as the main cutoff switch and prevents the combined branch circuits from drawing excess power into the house from the power company's line. Linked double breakers are needed for 240-volt circuits used for the kitchen range, hot-water heater, clothes drier and central air conditioner. Since each 240-volt circuit contains two hot wires, both breakers switch off simultaneously. Single large and small breakers protect 120-volt circuits (the smaller ones are simply spacesavers; they do the same work as the larger but allow two 120-volt breakers in the space of one). The empty space below the right-hand bank of breakers permits the addition of new circuits and their breakers.

Usually circuit locations and the breakers covering them are listed in a general fashion on the panel. For a more systematic way to identify which outlets are on which circuit, see page 20.

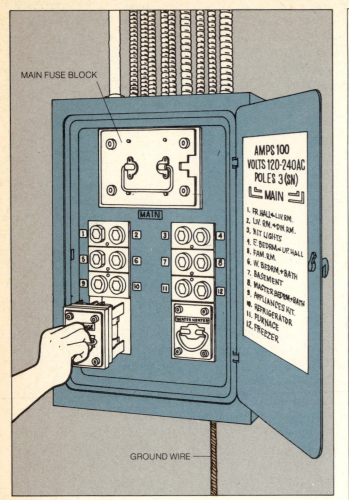

MAIN FUSE BLOCK

AMPS 100
VOLTS 120-240AC
POLES 3 (SN)
MAIN

1. FR. HALL + LIV. RM.
2. LIV. RM. + DIN. RM.
3. KIT LIGHTS
4. E. BEDRM + UP. HALL
5. FAM. RM.
6. W. BEDRM. + BATH
7. BASEMENT
8. MASTER BEDRM + BATH
9. APPLIANCES KIT.
10. REFRIGERATOR
11. FURNACE
12. FREEZER

MAIN

WATER HEATER

GROUND WIRE

Fuse panel. The typical fuse panel shown above has the main fuses—two knife-blade fuses—in the large pull-out block at the top of the panel. Removing this block turns off all power in the house. Some fuse panels have a cutoff lever that shuts off all the incoming power.

Below the main fuse block are plug fuses protecting 12 separate branch circuits. Those circuits for the general heating system and for the refrigerator, freezer and small appliances in the kitchen have 20-ampere fuses. The other branch circuits—for kitchen lights and for lights and appliances in other parts of the house—have 15-ampere fuses. Each of the two small pull-out blocks below the plug fuses contains a pair of ferrule-type cartridge fuses. One pair protects the 240-volt kitchen-range circuit and the other is for the 240-volt hot-water heater circuit.

Before removing any fuse, either to replace it or to work on the circuit, turn off all lights and devices on the circuit controlled by that fuse. To change a ferrule-type or knife-blade cartridge fuse, remove the appropriate block (*drawing, above*). For any other fuse, touch only the insulated outer rim to unscrew it. Do not stand in a damp spot or put your other hand on any object.

Types of Fuses

Plug fuse. The commonest fuse screws into the fuse panel like a light bulb. When excess current passes through the fuse, it overheats the metal strip in the center of the fuse, melting it at its narrowest, weakest point and opening the circuit. The appearance of a blown plug or similar screw-in fuse usually tells whether it blew because of a short circuit or an overload (*opposite*).

Time-delay. Like the plug fuse, the time-delay fuse screws into the fuse panel. However, its metal strip melts immediately only when there is a short circuit. In case of an overload, the strip softens but melting is delayed if the excess current is a momentary surge, as occurs when a motor-driven tool or appliance is first turned on. If the excess flow of current continues, the strip will melt and the fuse will blow.

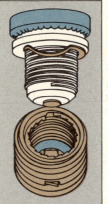

Type S. Although it functions like a time-delay fuse, a Type S fuse (also known as a non-tamperable fuse) cannot be screwed into the fuse panel unless an adapter base, shown in the lower part of the drawing, has been screwed in first. Because adapters are threaded for fuses of different capacities, it is impossible to use a fuse of the wrong rating. A 20-ampere Type S fuse, for example, will not fit a 15-ampere adapter.

In Canada, a Type C noninterchangeable fuse also can be used. This is not a time-delay fuse.

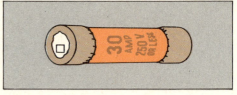

Cartridge fuses. Most cartridge fuses are the time-delay type. A blown cartridge fuse shows no visible sign of damage and must be checked with a continuity tester (*opposite, bottom*). The metal caps at the ends of a ferrule-type cartridge fuse (*drawing, above*) snap into spring-clip contacts in the fuse panel. Ferrule-type fuses are rated from 10 to 60 amperes and are generally used to protect separate circuits for individual large devices, such as kitchen ranges or shop tools. A cartridge fuse designed for more than 60 amperes, a so-called knife-blade fuse (*drawing, left*), has metal blades that snap into spring clips. In a home fuse panel, such fuses are generally used in the main connection between the incoming power line and the branch circuits.

Removing a cartridge fuse. Cartridge-type fuses are often used in auxiliary fuse boxes such as this one for a drier circuit. To replace the fuses, open the box by moving the cutoff lever to the "off" position. Using the correct type of fuse puller (pages 10-11), grasp the middle of the fuse and pull firmly to release it from the tight grip of the spring clips. If the circuit has just been in operation, be careful not to touch the metal end caps because they may be hot. Push the replacement fuse into the spring clips by hand. When replacing a fuse in a panel block, pull the block out, then use the fuse puller as above.

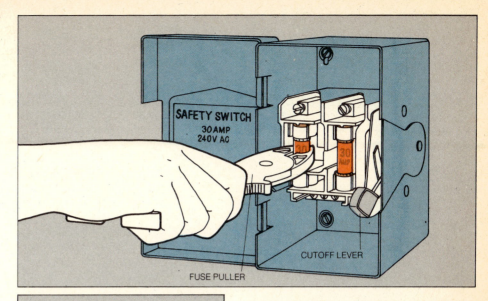

SAFETY SWITCH
30 AMP
240V AC

FUSE PULLER

CUTOFF LEVER

Detecting trouble. A glance at a blown screw-in fuse often reveals the cause of a circuit's failure. A short circuit melts the center of the metal strip rapidly, vaporizing the metal and leaving a discolored window (top drawing, right). An overload melts the strip but leaves a clear window (bottom drawing). If the cause of a blown fuse is uncertain from observation, see page 20 for the method of calculating an overload.

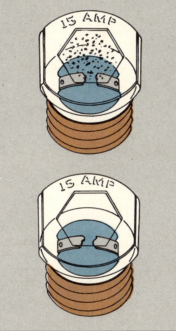

15 AMP

15 AMP

A Fused Receptacle

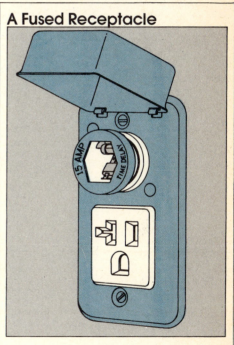

15 AMP   TIME DELAY

Testing a cartridge fuse. There is no external evidence of damage when a cartridge fuse blows, so if an appliance on a circuit protected by a cartridge fuse or fuses fails to operate, you have no way of knowing whether the fuse has blown or the appliance has simply failed. Remove the fuse and check it by touching a continuity tester (page 10) to the metal end caps. If the bulb lights, the fuse is good; the trouble is in the appliance. If the bulb does not light, the fuse is blown; replace it with a good one. If the appliance still does not work, call a repairman.

By replacing a standard receptacle with a fused receptacle, you can avoid having an entire circuit turned off by a device that might create an overload, blowing the fuse at the fuse panel. A fused receptacle is most useful in a workshop, where overheated power tools sometimes blow fuses. To install the fused receptacle, follow the instructions given on pages 54-55, but attach the black wire to the empty brass terminal on the back of the fuse. Fit the receptacle with a time-delay fuse to protect the circuit against the surge of current when a tool or appliance is first turned on.

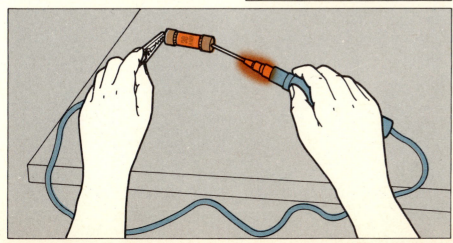

Grounding for Safety

In a normal electrical circuit, current flows to an appliance or light through a hot wire, and flows back to the service panel (and eventually the utility power station) through a white neutral wire. The hot wire is charged with voltage. The neutral one is at zero, the voltage of the earth—in fact, the neutral wire is connected to the earth at the service panel. Any deviation from this normal path is dangerous, and to protect you and your home against these hazards, electrical codes require a safety system called grounding, which keeps every outlet and cover plate at zero voltage.

Grounding guards against two hazards: fire and shock. The first can result from a short circuit, in which current leaks from a broken hot wire or connection by some path other than the normal one—across worn insulation, for example—and generates heat in the process. Because such a path offers high resistance, the circuit breaker will not be tripped.

A shock hazard generally arises when there is little or no leaking current, but the potential for abnormal current flow exists. If a bare hot wire touched the cover plate of a switch or a receptacle and the cover plate was not grounded, the voltage of the hot wire would charge the plate. If you then touched the charged plate, your body could provide a current path to zero voltage, and you could suffer a serious shock.

A grounding system uses a series of special wires that carry leaking current or abnormal voltages directly to a safe point of zero voltage, the earth. Because the electricity encounters little resistance in this special path, it is a high-amperage current—high enough to blow a fuse or trip off a circuit breaker. In effect, the grounding system ensures that if a circuit were to become seriously faulty, the circuit would no longer function.

All metal boxes and receptacles must be connected directly to the ground with a ground conductor, and some new switches are grounded as well (page 44). Plastic boxes, which do not conduct electricity, are not grounded, but the devices and cables contained inside them are. The method of grounding depends on the wiring that is used in your home. The ground wire in plastic-sheathed cable is normally the bare copper wire at the center of the cable, though ground wires insulated in green or green-and-yellow stripes are occasionally used. Some armored cable contains a bare aluminum or copper ground wire; in other styles of armored cable, the metal jacket itself is considered to be a satisfactory ground. In a metal conduit, the metal serves as a ground.

A ground connection must run without interruption from every receptacle and box on a circuit to the service panel. There it is connected to a metal strip called the neutral bus bar. The neutral wires from all the circuits—the normal return paths of current—are also connected to the bus bar. From a terminal on the bus bar a stranded copper ground wire provides the connection between the service panel and a metal pipe that passes into the earth (right).

Many appliances and power tools have their own ground wires to protect the user if a loose hot wire touches the metal housing, a situation as dangerous as a charged cover plate. A cord with a ground wire can be identified by its plug, which has three prongs. The plug should ideally be matched to a three-slot grounded receptacle, but in the United States (not in Canada) an adapter (opposite, top) can be used for a temporary match between a three-prong plug and a two-slot socket. To install a three-slot receptacle, see pages 54 and 55.

Wherever electricty is used in a damp location, a minor fault can cause a dangerous shock even if the circuit is properly grounded. For this reason, it is prudent to use a monitoring device called a ground-fault interrupter (GFI) in any damp or risky location. The GFI is usually required for receptacles in newly built bathrooms, garages, all outdoor installations, within six feet of the sink in kitchens, and in at least one location in basements. Check the local building code for other required locations. The GFI comes as a plug-in accessory for an existing receptacle, as a replacement receptacle (page 116), or as a circuit breaker installed in place of a standard breaker. Though expensive, ground-fault interrupters are worth the cost for the extra measure of safety they provide.

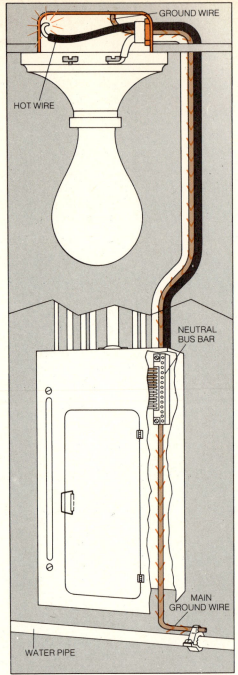

How a ground wire works. A hot wire in the fixture box above has come loose from its terminal and is touching the metal box. Because the wire is carrying 120 volts of electricity, the box is now dangerous to anyone who touches it. But the bare copper wire screwed to the box removes the danger by providing a path for current to the neutral bus bar in the service panel. From the bus bar, the current flows out through a ground wire to a metal water pipe that goes into the earth. And because the current is excessively high, a circuit breaker trips or a fuse blows, shutting down the entire circuit.

An alternate ground. In addition to a metal water pipe that goes at least 10 feet into the earth, modern electrical systems have a second, backup grounding means—usually a metal rod driven 8 feet into the soil. Two rods are used in houses with plastic water mains. Grounding systems need occasional checking by an electrician, for the soil may become too dry to conduct current, or the rod may corrode, particularly in acid soil.

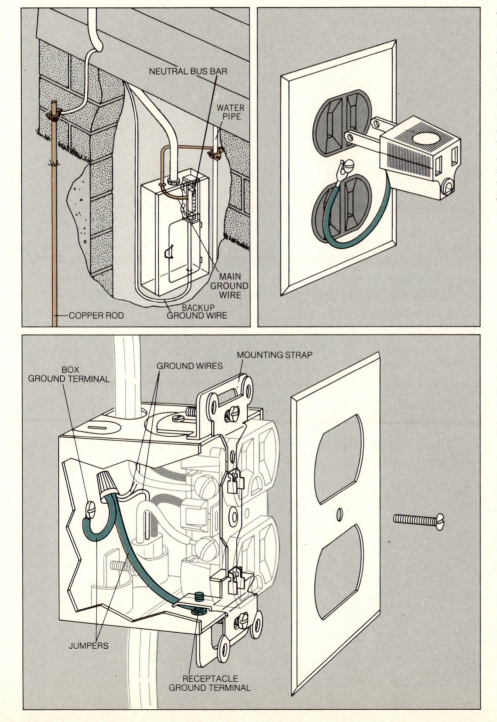

NEUTRAL BUS BAR

WATER PIPE

MAIN GROUND WIRE

BACKUP GROUND WIRE

COPPER ROD

BOX GROUND TERMINAL

GROUND WIRES

MOUNTING STRAP

JUMPERS

RECEPTACLE GROUND TERMINAL

A grounding-adapter plug. In the United States, you can accommodate a three-prong plug in a two-slot receptacle, with a grounding adapter. Check with a voltage tester to make sure that the receptacle cover-plate screw is properly grounded *(page 22)*. Then loosen the screw at the center of the plate, slip the brass connector of the green ground wire under it and tighten the screw. Since many cover-plate screws are short, recheck that the screw is grounded before plugging in the adapter. If it is not, replace it with a longer screw for a secure connection with the ground wire. In Canada, these adapters are not approved for sale.

How a receptacle's parts are grounded. All the metal parts of a receptacle and box that ought not carry current, but might if something went wrong, are given emergency routes to a ground through the bare copper wire of the cable. This wire is connected by a green jumper to the ground terminal on the mounting strap, the metal strip running through the inside of the receptacle. The mounting strap touches the box through its screws, and also is in contact with the metal cover plate through a screw. When a three-prong plug is plugged into the receptacle, it too contacts the mounting strap and thus grounds the device being used. A second jumper wire connects the ground terminal on the box to the cable's ground wire, providing another safe path to ground for current leakages.

In Canada, jumper wires and wire caps cannot be used on ground wires. Instead, the ground wires must be looped around a box ground terminal, and one ground wire must then be looped around a box terminal and connected to the ground terminal of the receptacle.

Mapping an Electrical System

Working on your house wiring is much simpler if you make a map to show which fuse or circuit breaker protects which receptacle, light and switch. A complete map enables you to pinpoint trouble, avoid overloads and know what parts of the house will be affected if you disconnect a circuit to work on it.

Before you start mapping, make sure that each fuse or circuit breaker at the service panel has a number. Then prepare your map. Use one sheet for each floor or even each room, indicating the location of all receptacles, lights and switches. Next, taking one room at a time, turn on all lights and small appliances such as radios or TVs. Do not turn on large appliances, such as a range or a clothes drier—each has its individual circuit, which can be identified later. If a receptacle is not in use, plug a small light into it; plug into both outlets of a duplex receptacle, since each one of them may be wired to a different circuit.

At the service panel, remove a fuse or trip a circuit breaker. Note its number and which lights or appliances went off. (An assistant to relay this information saves running back and forth.) On your map, note the fuse or circuit breaker number alongside the receptacles, lights and switches that stopped working. Then follow the same procedure for other rooms. Post map sections near the service panel in plastic envelopes that you can take with you when you work.

A guide to electrical work. In this plan of a bedroom wing, each outlet—receptacle, light fixture or switch (key)—is identified by the number assigned to the fuse or circuit breaker controlling it. Such a sketch tells which outlets will be affected if you turn off power to a circuit, and which fuse or circuit breaker to check when a circuit fails.

☐ SW	SWITCH
☀	LIGHT FIXTURE
⬚⬚	RECEPTACLE
→	SWITCH-TO-LIGHT LINK

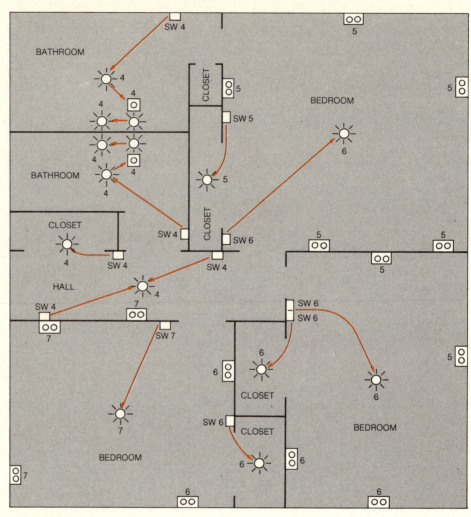

Calculating the Load on a Circuit

Plugging in an extra table lamp is unlikely to overload a circuit—such devices consume little electricity—but appliances like toasters and color TV sets are another matter. Before adding any heavy power user to a circuit, make sure it can bear the current it would carry.

List lights and other devices, including any additions, that will operate simultaneously on the circuit and total the wattage each consumes. This figure is printed on incandescent bulbs and fluorescent tubes and on data plates of appliances. Then divide the total wattage by 120 to get the number of amperes of current. The total amperage must not exceed the circuit capacity, which is marked on the fuse or circuit breaker. The same calculation tells if an overload caused a tripped breaker or a blown fuse. Total the wattages of all devices in operation, and calculate amperage as above to see if the circuit capacity was exceeded.

How to Check Your Work

Two inexpensive devices—a voltage tester and a continuity tester—are all you need to check any wiring connections described in this book. The voltage tester is primarily a safeguard against shock, used to make absolutely sure that there is no voltage in the circuit you are working on. The continuity tester helps diagnose electrical faults.

The voltage tester has no source of power built into it; it lights when its probes are touched to anything that is charged with electricity. The probes are designed to fit into the two slots of a receptacle, making it possible to check whether the power is off or on without first removing the cover plate. In some instances, the voltage tester is used with the power on, to locate the feed cable bringing electricity from the service panel. The most useful tester is designed for about 90 to 500 volts.

The continuity tester (page 23) has its own power source, a small battery that lights a bulb when a continuous path for current lies between the probe and alligator clip. The device must be used only when power to an appliance or a circuit is turned off. The alligator clip is attached to one point and the probe touched to another; if the bulb does not light, there is a break in the path of the current between the two points.

By this method, a continuity tester can detect interruptions in a circuit caused by a break in a cord, a bad connection, or a defective socket or switch. It can also check the soundness of a cartridge fuse (page 17) and the operation of a toggle switch (page 44). When the tester is not in use, insert the probe in the plastic sleeve provided for it; otherwise, the probe may accidentally touch the alligator clip, lighting the bulb and wearing out the battery.

Using a Voltage Tester

Testing a receptacle. After removing the cover plate of a receptacle, set the prongs of the voltage tester against the bare ends of the black and white wires at the points where they are attached to the receptacle. Be sure to test both sets of wires on a receptacle that, like this one, has wires at all four terminals. The bulb should not light during the test, indicating that the power is indeed off. If it does light, return to the service panel to find the fuse or circuit breaker that controls the receptacle and turn off the circuit. A switch is tested in a similar manner (page 43).

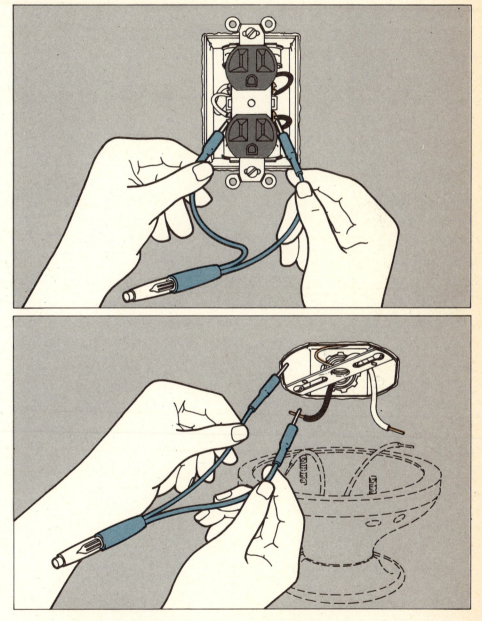

Testing for voltage at a light. With a good bulb in the fixture, turn the light on. Then turn off the circuit breaker: the extinguished light will show that you have turned off the correct one. Turn the wall switch to the "off" position. Remove the screws or nuts that hold the fixture to the ceiling or wall, and pull the fixture out of the box to expose the wires. Remove the wire caps with one hand; hold the fixture with the other. Grasp the wires by the insulation only and untwist them while supporting the fixture. Set the fixture aside. Set one probe of the voltage tester on the bare end of the black cable wire and the other against the metal box, which is grounded. Then test from the same black wire to the white cable wire. Finally, test from the white cable wire to the grounded box. The tester bulb will not light during these tests if the power is indeed off.

Checking a receptacle's ground. Test the grounding of a newly installed receptacle by inserting one probe of the voltage tester into the semicircular ground slot and the other into each of the elongated slots successively. The tester should light when the probe is plugged into the hot slot (in modern receptacles, this slot is slightly shorter than the other slot). If neither slot lights the tester, the receptacle is not grounded and the wiring must be corrected.

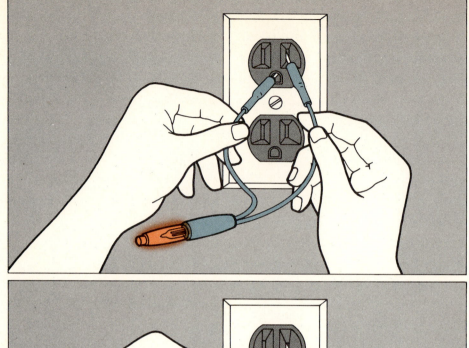

Checking the cover-plate ground. Before installing a grounding adapter plug (*page 19*) for use with an older receptacle that has no ground slot, check to be sure that the cover plate is properly grounded. Set one probe of the voltage tester against the mounting screw of the plate and insert the other probe into each of the straight slots successively. The tester bulb should light when the second probe is placed in the hot slot. If it does not light in either slot, the cover plate is not grounded and the wiring must be corrected. The same test should also be carried out on any newly installed receptacle to make sure that its cover plate is grounded.

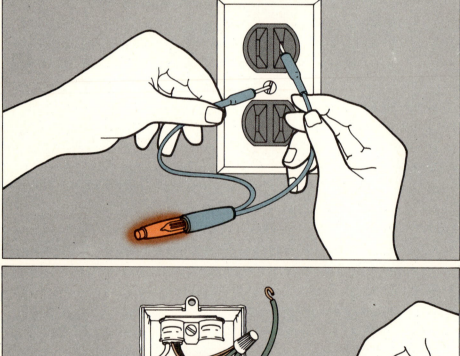

Testing for incoming power. Occasionally you will need to know which of two or more black wires is the feed connecting a box to the house service panel. Turn off the power, pull the device from the box, disconnect the black wires and pull the wires apart so that they are not touching one another or any other equipment.

While you hold one probe of the voltage tester against a black wire and the other against the grounded metal box, call to a helper to turn the power back on at the service panel.
Caution: Be very careful not to touch the wires or the box with your hands, and do not push the black wire against the box. If the tester lights, you have located the incoming wire. If it does not light, carefully try another black wire. When you find the feed wire, have your helper turn off the power and use the voltage tester to make sure that he has done so. Identify the feed wire by marking it with tape. (If the outlet box is plastic, test between a black wire and the ground wire.)

Using a Continuity Tester

1 Checking a lamp socket. When a bulb is good but a lamp does not work, the fault may be an open circuit in the socket. Unplug the lamp, remove the bulb and take the socket apart (*page 30*). Then clamp the alligator clip of a continuity tester to the threaded metal tube and touch the probe to the silver-colored, or neutral, screw terminal. The threaded tube and the terminal should both be neutral and the tester should light; if it does not, the socket has an open circuit and should be replaced.

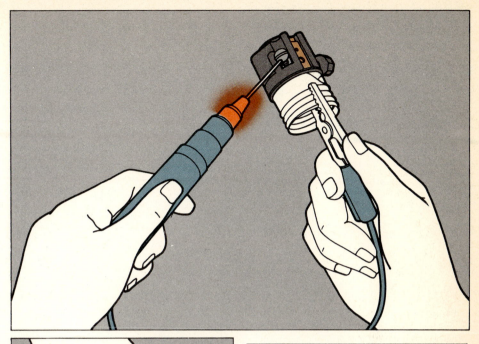

2 Checking a lamp switch. Clamp the alligator clip to the brass, or hot, screw terminal and touch the probe to the rounded contact tab at the center of the socket. Turn the switch off, then on. If the switch is faulty—a common problem in lamp sockets—the tester will not light; replace the socket with a new one. If the tester lights when the switch is on, but the socket still does not work with a bulb you are sure is good, the problem may be that the contact tab does not make proper contact with the bulb. Raise the free end of the tab slightly with the screwdriver tip. If the lamp still does not light, the fault is not in the switch; check the cord and plug (*pages 30-31*).

Check the switch of a three-way socket (*right*) in the following sequence. Clamp the alligator clip on the brass screw terminal. Test the four switch positions, touching the probe to the small vertical tab in the base of the socket and then to the rounded tab as you turn the switch. In the first "on" position, the tester should light when the probe touches the vertical tab but not when it touches the rounded tab. In the second "on" position, the tester should light only when the probe is touched to the rounded tab. In the third "on" position, the tester should light at either tab, and in the fourth position—"off"—at neither tab.

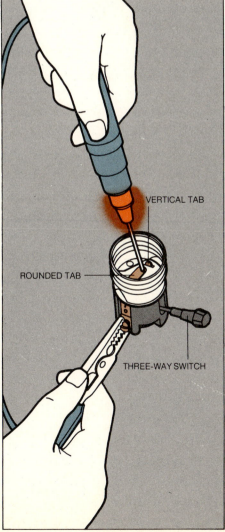

VERTICAL TAB

ROUNDED TAB

THREE-WAY SWITCH

Tips for Safety

You are protected against electrical shock while working on house wiring if you follow the basic precautions below. To ensure the long-term safety of the work after it is done, follow standard procedures exactly—take no shortcuts—and test the finished job.

☐ Turn off the power in the circuit you plan to work on by switching off the circuit breaker or removing the fuse that protects that circuit.

☐ Test the wires in a box with a voltage tester to make sure the power really is off. Test from the black wires to the grounded box and the white wires, and test from the white wires to the box.

☐ Never touch parts of the plumbing system or gas piping when you work with electricity—or while you use any electrical appliance.

☐ Do not stand on a damp floor when you work with wiring.

☐ Unplug a lamp or appliance before working on it. Pull on the plug itself; do not pull on the cord.

☐ Check your work after installation with the power on. Use a voltage tester to check from the black wire to the grounded box and to the white wire; the tester should light. Check from the white wire to the grounded box; the tester should not light.

Wires, Cables and Conduit

Electricity is conducted along house circuits by wires grouped together in the form of cable or, in some cases, contained in conduit. Cable consists of a preassembled combination of wires within a protective outer sheath made of metal or plastic. Conduit is simply piping of steel or plastic through which several wires are threaded after the pipe has been installed.

The hot and neutral wires that conduct electricity in both cable and conduit are individually insulated. The ground wire is not insulated in cable. Metal—but not plastic—conduit needs no ground wire, since the pipe is its own ground.

With few exceptions, the wire used for home circuits is Type T, so called for its thermoplastic insulation, which is capable of withstanding a wide range of temperatures. Type TW, a weatherproof variant, is used outdoors or in a damp location, such as a basement.

The type and size of conductor wires provided in metal-sheathed cable and of those intended for installation in conduit are marked on the insulation *(right)*; in plastic-clad cable, wire type and size is indicated on the plastic sheathing *(bottom right)*. The wires themselves are copper. Wires are color-coded to avoid error in connecting them to terminals. The hot wire is almost always covered in black insulation; the neutral wire is always covered in white or gray. The ground wire is left bare, wrapped in paper, or, in some fixtures, switches and receptacles, insulated in green. If there are other hot wires—for three- or four-way switches *(pages 46-47)* or for 240-volt circuits *(page 13)*—they are coded red or blue.

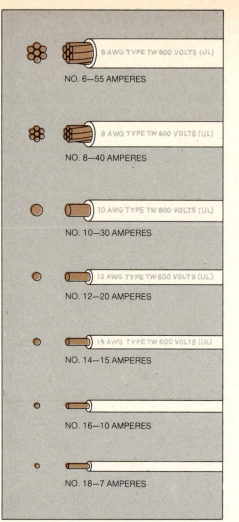

NO. 6—55 AMPERES

NO. 8—40 AMPERES

NO. 10—30 AMPERES

NO. 12—20 AMPERES

NO. 14—15 AMPERES

NO. 16—10 AMPERES

NO. 18—7 AMPERES

Wires for the house. The wires used in house circuits are shown at left. On larger sizes, wire diameter is indicated by a gauge number printed on the insulation and based on the American Wire Gauge system (AWG) —the smaller the number, the greater the diameter. Wire type is also marked. The smaller sizes—No. 18 and No. 16—are too small for printed identification.

Wire sizes No. 14 and No. 12 are used in standard 120-volt circuits for lighting and receptacles for TVs, clocks and other small appliances. Size No. 10 is used in 120- or 240-volt circuits carrying electricity to such major appliances as space heaters and clothes driers. Wire sizes No. 8 and No. 6 are the most widely used in 240-volt circuits for electric ranges and central air conditioners. Wire sizes No. 18 and No. 16 are reserved for low-voltage systems like bells and intercoms; stranded, they are also used for lamp cord.

The maximum current a wire of a given diameter can safely carry, stated in amperes, is the wire's ampacity. The smaller the wire, the greater its resistance to the flow of current and the greater the heat-generating friction, which could destroy insulation and even kindle a fire; thus large currents require large wires. The ampacities at left are for copper wires, the commonest kind in house circuits. Aluminum wire, which does not conduct electricity as efficiently as copper, has an ampacity approximately equal to that of copper wire two sizes smaller; for example, aluminum wire No. 12 has about the same ampacity as copper wire No. 14.

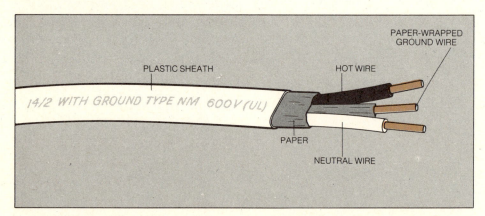

Plastic-sheathed cable. Nonmetallic cable— one brand name is Romex—is the commonest form of wiring in houses built since 1960. Least expensive, it is also the easiest to install because of its flexibility. The insulated hot and neutral wires and the uninsulated ground wire are enclosed in a protective sheath, generally of plastic. Not all cable has a ground wire, and for safety's sake cable without one should not be used.

Three kinds of nonmetallic cable are usually needed in house wiring: Type NM, for ordinary indoor wiring; Type NMC, for damp indoor locations and for outdoor wiring above-ground; and Type UF, for buried outdoor circuits. The sheathing is marked for *(left to right)* wire size, number of conductors, presence of a ground wire and cable type. Shown here is Type NM 14-gauge, two-conductor, grounded indoor cable.

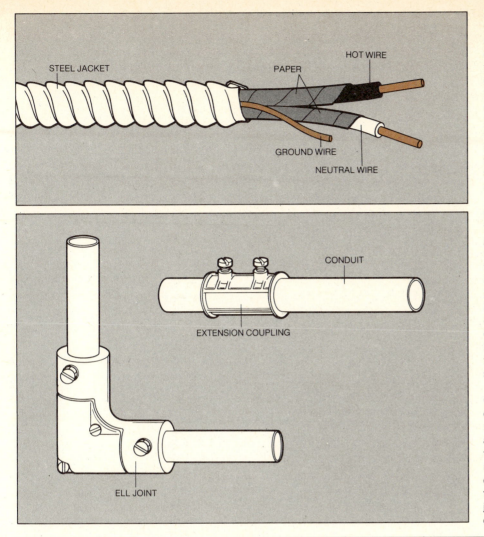

STEEL JACKET   PAPER   HOT WIRE

GROUND WIRE   NEUTRAL WIRE

EXTENSION COUPLING   CONDUIT

ELL JOINT

Flexible armored cable. In armored cable, the wires have a jacket of galvanized steel, spiral-wound for flexibility. Each insulated conductor wire also has a paper wrapping so the insulation will not be damaged when the cable is flexed; the ground wire, if any, is bare. Because the steel jacket is subject to corrosion, armored cable—known as BX — may be used only in dry indoor locations. Often found in older homes, it is still required by some local codes. It is also recommended for use where wires need extra protection from damage— as within a wall space where nails are likely to penetrate.

Conduit. Of the many types used to hold wires, the conduit most often seen in homes is a galvanized steel pipe called thin-wall conduit, which protects wires from damage better than armored or plastic-sheathed cable. Once conduit is installed, extra wires may be threaded through *(page 114)*. As many lengths of conduit as needed can be joined to each other or to cable by connectors like the extension coupling and "ell" corner joint at left. Since thin-wall conduit does not have the flexibility of cable, though, it cannot be snaked behind walls; it is difficult to install.

Other forms of conduit, sometimes preferred outdoors, include rigid conduit, which is a more durable metal pipe *(pages 114-115)*; and PVC (polyvinyl chloride), rigid plastic tubing that resists corrosion better than metal. PVC is less expensive than metal conduit, but it can be a hazard: it emits toxic fumes when burning. Also, PVC must be joined to connectors with adhesive—and cannot be taken apart again.

The Hazards of Aluminum Wiring

Because of a shortage of copper wire in the 1960s, some house wiring installed after 1965 is either copper-clad aluminum or pure aluminum. Copper-clad wire is safe, but pure aluminum wire, installed in some two million homes, can be dangerous: The U.S. Consumer Product Safety Commission attributes 500 fires to its use. Manufacturers modified aluminum wiring in 1972 to make it safer, but a few homes and additions built later were still wired with the old cables and devices.

The problem lies in two chemical reactions that occur on the surface of aluminum. One involves the corrosion that takes place when dissimilar metals meet—as between the aluminum wire and standard copper-alloy outlet terminals. In the other reaction, aluminum

wire oxidizes soon after its insulation is removed, exposing it to air. Either reaction increases resistance to current and thereby generates dangerous heat.

Both reactions can be prevented if the wires are installed scrupulously, using an antioxidizing paste and the improved switches and receptacles introduced in 1972. These devices, marked CO-ALR, have terminals designed to prevent contact between dissimilar metals.

If your wiring is pure aluminum (you can tell by the AL markings on cable sheaths in unfinished areas), check your receptacles and switches to make sure they are marked CO-ALR. If not, the commission recommends that house wiring be completely redone or modified throughout with COPALUM connectors—special devices sold only to electricians

schooled in their use. These remedies are costly, but they are the only foolproof methods of ensuring safety.

Electrical inspectors, however, point out that aluminum wiring is still legal in most areas, and is still used widely in high-amperage circuits and commercial buildings. If you have an aluminum-wired system in your home and you elect to live with it, have it checked periodically by an electrician knowledgeable about its hazards. Avoid plugging in high-wattage appliances in rarely occupied rooms. Be especially alert for danger signals: warm cover plates, switches or receptacles; mysteriously inoperative devices; odors; or smoke. If you notice any of these, call an electrician. Do not try to repair or improve an aluminum-wired system yourself.

Working with Wire

In house wiring, there are two kinds of connections: wires are either attached to terminals such as those on switches or receptacles, or they are spliced to other wires. For safety reasons, a wire splice must be made only inside an outlet or junction box—never in the wires running between the boxes. Before attempting any wire connections, shut off power to the circuit. Remove plastic sheathing from the cable *(right)* and prepare the end of the wire by stripping off the insulation *(bottom right)*.

To attach the stripped wires to a receptacle or switch, either wrap the wire around a terminal screw or insert it in a self-gripping slot, depending on which type of terminal the device has *(opposite, top)*. Only one wire can be attached to each screw or push-in terminal. If you need to connect more than one wire—as may be the case with ground wires *(page 19)* or switch-controlled receptacles *(pages 99-100)*—use a jumper wire. A jumper is an extra piece of wire of the same diameter and with the same color of insulation as the wires to which it is attached. Strip both ends of the jumper, splice one end to the other wires with a wire cap and attach the other end to the terminal. The jumper should just reach from the wire cap to the terminal.

To make splices, use a pair of pliers —twist the stripped wires around each other, then secure them with a wire cap. The most practical type of wire cap is the one shown on the opposite page. It consists of a cone-shaped insulating shell of hard plastic with a spiral of copper inside. When the cap is twisted over the ends of bared wires, the copper spiral grips the wires tightly together. Another type has a rounded shell of soft, instead of hard, plastic. It is less susceptible to breakage, but unlike the hard-shelled type, it cannot be removed. To correct a faulty splice or to add a new wire, snip the wires below the cap and start again.

Wire caps are manufactured in several sizes. To determine which size is appropriate for the job, consult the table on the container; it will specify the wire size and the number of wires that the cap is designed for.

Stripping Off Insulation

Removing sheathing. Place the cable on a steady surface and with your thumb indicate on the cable a distance of about 8 inches from the end. Insert a utility knife into the sheathing near your thumb and cut down the middle of the cable a shallow groove to serve as a guide for cutting the sheath. Then follow the groove and cut as deep as the ground wire *(below, left)*. Be sure to avoid touching the insulated wires with the knife; if you mar the copper conductors, cut away the cable and begin again. Peel back the plastic sheath to the beginning of the cut, tear off the paper that is wrapped around the wires and snip off the loose section of sheathing *(below)*.

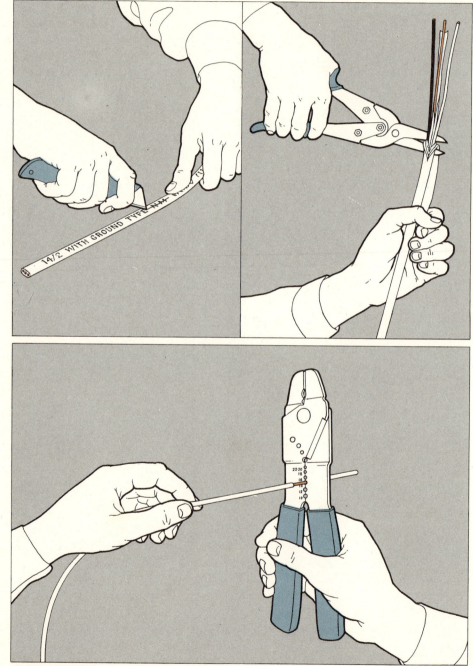

Stripping insulation. Remove wire insulation with a special stripping tool or with the part of an electrician's multipurpose tool *(above)* that is designed for that purpose. Place the wire in the hole sized for the wire, close the tool over the wire and twist the tool back and forth until the insulation is cut through and you are able to pull it free. Do not attempt to strip insulation with a knife. It endangers fingers and generally nicks the wire metal, creating an electrical hazard.

Two Kinds of Terminals

Screw terminal. Strip only enough insulation to allow the bared end of the wire to be wrapped three quarters of the way around the terminal screw. With long-nose pliers, twist the bare end into a loop, loosen the terminal screw, and hook the loop clockwise around the screw so that when the screw is tightened it will help close the loop.

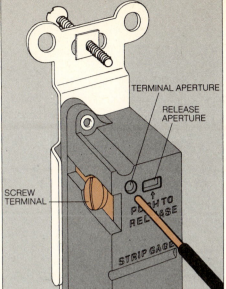

TERMINAL APERTURE

RELEASE APERTURE

SCREW TERMINAL

PUSH TO RELEASE

STRIP GAGE

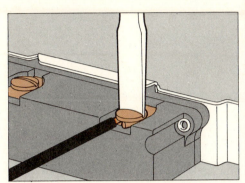

Push-in terminal. Some switches and receptacles are designed so that you can loop the wire around a screw terminal or insert it through an aperture that automatically grips for a solid connection. To attach a wire to such a push-in terminal, strip insulation as indicated by the strip gauge marked on the device—usually about ½ inch. Insert the bare wire into the terminal aperture (drawing, left), and push it in up to the insulation; a spring lock will grip the wire in place. To remove a wire from such a device, insert the blade of a small screwdriver or the end of a stiff piece of wire into the release aperture next to the push-in aperture, press the spring lock and pull the wire free.

Making Connections with Wire Caps

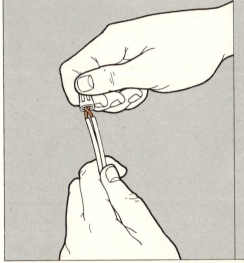

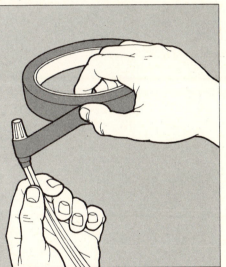

Solid wires. Strip about ¾ inch of insulation from the wires to be joined and hold them parallel. Slip a wire cap over the bare wire ends; twist the cap clockwise around the wires, pushing them hard into the cap (far left). If any bare wire remains exposed, remove the cap, cut the wires to the proper length, then twist the wire cap back on.

To make certain that the wire cap will not jar loose when the wires are pushed back into the box, many electricians secure the cap with insulating tape (left). Wrap the tape around the base of the cap, then once or twice around the wires and finally around the base of the cap again.

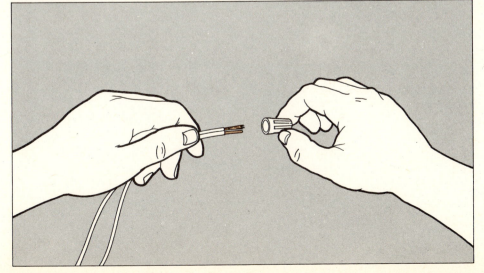

Stranded wire to solid wire. Strip about ¾ inch of insulation from the solid wire and about ¾ inch from the stranded wire. Hold the two bared ends parallel, push a wire cap over the two of them and twist the cap clockwise until it is tight. Then wrap the wires and cap as above.

Replacements that Modernize

Paint peels, roofing deteriorates—but the wiring system of your home seems to go on forever. And indeed it should. The cables and outlet boxes that form the heart of the system are nearly age-proof. Nestled safely behind walls and ceilings, subject to almost no strain or movement, they normally last as long as the house itself.

The parts mounted in outlet boxes—the switches, lights and receptacles that put the system to work—are subject to wear and eventually fail. They cannot be repaired, but they are inexpensive and easy to replace. Even though the cause of a breakdown may seem mysterious, it can usually be detected by following a methodical, common-sense procedure of diagnosis.

If a plug constantly drops out of a receptacle, the fault might lie out of sight, in receptacle contacts that have lost their tension. But before you pull the receptacle out of the wall, look closely at the plug, which is just as likely to be defective and is even easier to replace. Elsewhere, you may have to go through a series of steps before you can locate the trouble. When the flick of a switch does not turn on a light, the obvious first step is to check the light bulb. If it proves sound, you should go on to the fuse or circuit breaker, which is easy to replace or reset. The next step—if you have not found the trouble—poses a question: Should you first remove the wall switch or dismantle the ceiling light? Since switches have moving parts, they are more likely to break down than light fixtures; what is more, it is easier to work on a wall than a ceiling. Only after this elimination process would you have to climb a ladder to look for a broken light socket or a loose connection. If the fixture is the fluorescent type, which contains fairly complex components and comes in several varieties, you might continue your diagnosis by using the troubleshooting chart on page 39 to help identify the problem.

The most gratifying replacements, however, are made not because of trouble-caused necessity but because of an opportunity for improvement. Long before a part breaks down, you may replace it to better your wiring system. There is good reason to convert an incandescent light fixture to a fluorescent simply because fluorescent lights are more efficient and reduce electric bills. New switches and receptacles can be safer than old ones because they have locking or tamperproof features (pages 42, 57). Many are better looking and, like the timer switch on the opposite page, some can serve special functions. Replacement parts can even increase the versatility of your wiring system without the addition of a single cable or outlet box. To get a new receptacle without new wiring, for instance, you can replace a wall light with a light-and-receptacle fixture or, in certain situations, a wall switch with a switch-receptacle (page 55).

How to Rewire a Lamp

Despite the great variety of lamp shapes and sizes, the electrical components and the way they are wired together are much the same in all. The components consist of one or more sockets, a switch, a plug and a cord, all of which can easily be replaced. Usually the supporting skeleton is a threaded pipe, invisible in the center, that holds socket, lamp body and base together, and also guides the cord from the base to the socket. The socket cap may fasten the lamp body to the pipe. If you take the cap off, make sure the lamp is supported while you work on it or the base and body may slip apart.

The part of the lamp that most often fails is the switch. If it is a separate unit in the base, replace it with a new one. Most lamps, however, have switches in their sockets (which also may need replacement if they corrode). Standard sockets are almost always interchangeable, although they may use either a pull chain, push button or rotating switch. Even the special socket used with a three-way bulb is hooked up in exactly the same way as the standard one-way socket shown below. Rewiring a lamp fitted with two sockets (opposite) involves a few additional connections.

A plug should be replaced if its casing is cracked or if its prongs are so loose that they no longer make a good connection at the receptacle. Use polarized plugs (page 52, first picture) if they fit the receptacles in your home; by wiring them correctly, you will be sure that the screw shell of the lamp is always neutral, not hot (page 23). In a standard plug (right) the cord wires must first be stripped of insulation, tied in an Underwriters' knot and then screwed to the plug's terminals. Other plugs clamp directly to the cord, eliminating the need to strip the wires or screw them down. However, because no Underwriters' knot or terminal screws are used in clamp-type plugs, their connection to the wire is not as strong as in the standard plug. They are not recommended for lamps that are unplugged often.

If a lamp cord is damaged, buy a new cord instead of splicing the old one—a safe splice is more work than it is worth, and sloppy splices are a hazard. When buying a new cord, specify SPT 16- or 18-gauge lamp cord, called rip cord because the insulation is grooved between the wires so they can be pulled apart.

The Simplest Lamp: A Switch-and-Socket Combination

1 **Socket.** Be sure to unplug the lamp and remove the light bulb (below) before starting repairs. On the outer shell of the socket is a point marked "press"; push against this point with a screwdriver to separate it from the socket cap. Pull off the shell and the insulating sleeve. Remove the socket assembly by loosening the terminal screws and taking the cord wires off the terminals.
If only the socket is being replaced and the socket cap is not bent or corroded, leave it in place and reuse it, attaching only the new socket, sleeve and shell as explained in Step 4. If the socket cap is damaged, untie the knot in the cord, loosen the setscrew in the cap, unscrew it from the top of the center pipe and replace it.

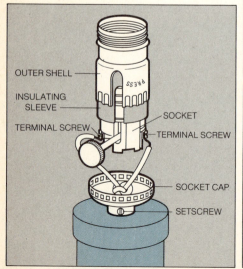

OUTER SHELL
INSULATING SLEEVE
TERMINAL SCREW
SOCKET
TERMINAL SCREW
SOCKET CAP
SETSCREW

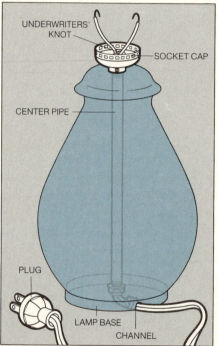

UNDERWRITERS' KNOT
SOCKET CAP
CENTER PIPE
PLUG
LAMP BASE
CHANNEL

2 **Putting in a new cord.** Untie the knot in the end of the cord resting in the socket cap and pull the old cord out through the center pipe. Untie the knot in the cord at the channel in the bottom of the lamp and pull it free of the lamp. If the old plug is reusable, save it to attach to the new cord according to the procedure in Step 3. Push the new cord through the channel in the lamp base and knot it just inside. Make sure you will have enough cord to reach the socket cap with about 4 inches to spare. Snake the cord up through the center pipe and work it through the socket cap.

3 **Tying an Underwriters' knot.** Separate the two wires of the cord you have pulled up through the socket cap, parting them gently 2 inches along the grooved insulation. Strip ½ inch of insulation from the end of each (page 26) and tie the cord ends into an Underwriters' knot (below). Tighten the knot and pull the cord back into the center tube so that the knot nestles in the neck of the socket cap. If you are installing a plug with screw terminals, insert the cord end through the neck of the plug and tie a second Underwriters' knot. After tying it, pull back on the cord so that the knot rests in the neck of the plug.

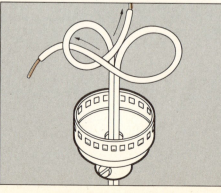

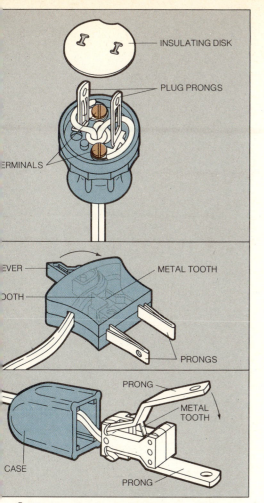

INSULATING DISK

PLUG PRONGS

TERMINALS

LEVER

METAL TOOTH

TOOTH

PRONGS

PRONG

METAL TOOTH

CASE

PRONG

PRONG

4 **Attaching the plug and socket wires.** The handiest plugs to use on a lamp are the quick-clamp plugs shown here, which do not require you to strip the wires. With quick-clamp plugs, either a lever forces the cord onto a tooth at the top of each prong (*top*) or each prong swings in until its tooth pierces the cord (*bottom*). If you use a polarized plug, make sure the hot wire connects with the tooth for the narrow prong and the neutral wire with the tooth for the wide prong (*page 23*). The plug case then slides over the plug body and holds the prongs in place.

At the socket cap, loop the stripped and twisted tips of wire around the terminal screws of the new socket. The hot wire should go to the brass terminal and the neutral wire to the silver-colored terminal (*page 23*) to ensure that the threaded tube is neutral and the rounded tab is hot. Tighten the terminal screws, then fit the new shell and sleeve over the socket and push them down until they snap into the socket cap.

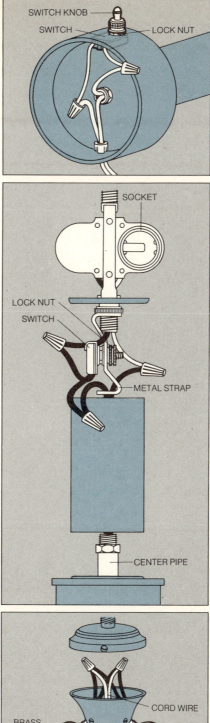

SWITCH KNOB

SWITCH

LOCK NUT

SOCKET

LOCK NUT

SWITCH

METAL STRAP

CENTER PIPE

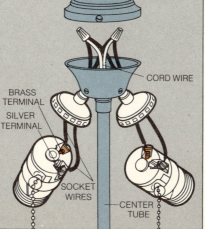

CORD WIRE

BRASS TERMINAL

SILVER TERMINAL

SOCKET WIRES

CENTER TUBE

Separate Switches and Twin Sockets

One socket with a separate switch. To replace a separate switch mounted in the base of a single-socket lamp, remove the knob and unscrew the lock nut. If the lamp base is sealed, remove the bottom cover, pull out the switch and remove the wire caps holding switch wires to the cord. Reattach the wires of the new switch to the cord, place the switch in its mounting hole in the base and reverse the disassembly procedure.

Two sockets with a separate switch. In this arrangement, a single assembly includes both sockets. One black wire and one white wire serve both sockets and are a permanent part of the unit. A single switch located below the sockets controls both. To replace the socket assembly or the switch, first remove the switch knob. Then remove the socket, switch and wiring by turning the socket assembly counterclockwise until it is free of the center pipe. Remove the wire caps holding the two socket wires. Unscrew the socket from the metal strap and remove it. Reverse the procedure to put in the new socket.

To replace only the switch, remove the socket and switch assembly. Unscrew the lock nut holding the switch to the strap and pull the switch from the strap. Disconnect the old switch wires from the cord and socket wire, reattach them to the new switch and refasten with the lock nut.

To remove only the cord, unscrew the wire caps holding the cord wire to the white wire and the other cord wire to the switch wire. Reverse the procedure to attach the new cord. If the plug is polarized, connect the hot wire to the switch. All other steps are the same as for a single-socket lamp.

Two switch-and-socket combinations. In this lamp, as in the single-socket lamp opposite, each socket is served by two wires and each has its own switch. The socket wires branch off from the main cord at the top of the lamp's center pipe. Replace the sockets, following the same procedure that was used for a single-socket lamp.

To replace only the cord, remove the cover at the top of the lamp and unscrew the wire caps holding the socket wires to the cord wire. Replace the cord wire using the same method as for a single-socket lamp, but after stripping ½ inch of insulation from the separated cord ends, attach the socket wires to the cord wires with wire caps. Be sure to attach the wires from the brass-colored socket terminals to one of the cord wires and the wires from the silver-colored socket terminals to the other cord wire.

Changing Fixtures: A Quick Step toward a Restyled Room

One of the easiest ways to transform the appearance of a room is to take out an outmoded light fixture and put a more stylish one in its stead. The replacement job—on ceiling or wall—consists of three operations: removing the old fixture, wiring the new one, and then mounting it to the box. Wiring is essentially the same for all fixtures, whether they use incandescent bulbs or fluorescent tubes, but the hardware needed for mounting varies. These variations, which apply to most wall and ceiling fixtures, are shown at right and below. The method used for hanging a chandelier is shown on page 34, and the methods for hanging fluorescent fixtures on page 37.

Before starting the job, turn off the power to the fixture circuit at the service panel. Merely turning off the wall switch is not sufficient, since in certain situations (page 13) the switch may not turn off power to all wires in the box.

With the power off, remove the screws or nuts holding the fixture in place, pull it away from the box and hang it from the box with a hook made from a wire coat hanger. Do not let the fixture hang by the wires; a dangling fixture may pull the wire connections loose before you are ready to remove them. Detach the connections to the fixture. Do not tamper with any of the other wires. Strip ½ inch of the insulation from the wires of the new fixture and then connect them to the house wires in the box, black to black, white to white.

If the new fixture is the same size, shape and weight as the one being replaced, you can probably reuse the old hardware in mounting it. If a different-sized fixture is being installed, however, you may need other hardware. Either remove the old fixture first, examine the box and buy the necessary parts, or buy a package of parts in assorted sizes.

Mounting a simple fixture. The single-bulb fixture, usually made of porcelain and often used in garages and basements, has no wiring of its own. The holes in its canopy are spaced so that they line up with the screw holes in the box tabs for direct mounting. First attach the black house wire to the brass-colored screw terminal, and then the white house wire to the silver-colored terminal. Finish the job by folding the wires neatly into the box, and then screw the fixture on.

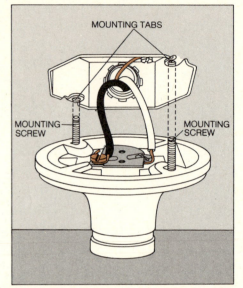

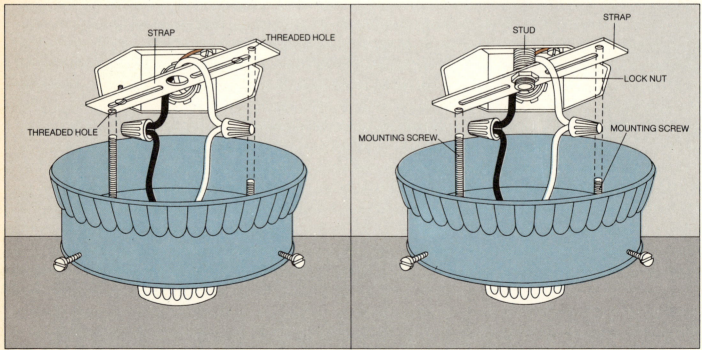

Strap-mounting a fixture to a stud. If the base of a fixture is so large that the canopy screw holes will not line up with the box tab holes, the fixture must be mounted with an adapter called a strap. If the box has a stud projecting from its middle, thread the strap's center hole onto the stud and secure it with a lock nut (above, right). For a studless box, insert screws through the strap slots into the box tabs. Connect the fixture and wires to the house wires with wire caps—black to black and white to white—and fold them into the box. Screw the fixture to the holes in the strap.

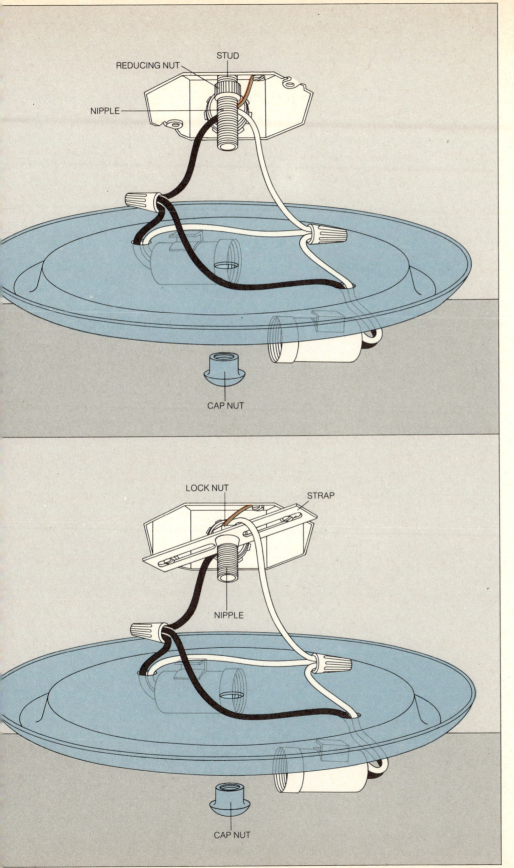

STUD

REDUCING NUT

NIPPLE

CAP NUT

LOCK NUT

STRAP

NIPPLE

CAP NUT

Securing a center-mounted fixture. Attach to the stud a reducing nut—one threaded to fit the stud at one end but threaded differently at the other end to fit a smaller threaded metal tube, called a nipple (*left*). Insert the nipple into the reducing nut and screw it in tightly. If the box has no stud, fasten the nipple into the center hole of a strap and screw the strap to the box tabs (*bottom*). Connect the two white wires from the sockets of the new fixture to the white house wire. In the same way, connect the two black socket wires to the black house wire. Place the fixture over the box so that the nipple protrudes through the center hole of the fixture. Then fasten the fixture to the nipple with a cap nut.

Chandeliers

A chandelier that weighs less than 10 pounds can be attached to the tabs of a ceiling box with a strap and nipple *(page 33)*. But a chandelier of 10 pounds or more requires some special hardware and can be safely attached only to a ceiling box equipped with a stud *(right)*. Therefore, if you want to replace a lightweight chandelier with a heavier one and the present ceiling box has no stud, you will have to substitute a new box with a stud. To remove the old box, shut off the power to that circuit and follow the instructions on page 32. Install the new box as directed on pages 70-75.

Rewiring a chandelier is similar to rewiring a two-socket lamp *(page 31)*, but the number of steps varies, depending on the number of sockets and whether the sockets are prewired or not. Both types of sockets are shown on the opposite page. With either type, however, the basic connections are the same. Each individual socket is served by two branch wires. The branch wires from the hot, or brass-colored, terminal on each of the sockets are attached with a wire cap to one of the chandelier's two main-cord wires. The branch wires from all the neutral, or silver-colored, socket terminals are attached to the other main-cord wire. Then the main wires are connected to the house wiring, hot wire to the black house wire and neutral wire to the white house wire. Use a continuity tester—with added wire from the alligator clip to a connection, if necessary—to check your work.

Never let a chandelier hang by its wires. Either have a helper hold the chandelier or suspend it from the box with a hook made from a wire coat hanger.

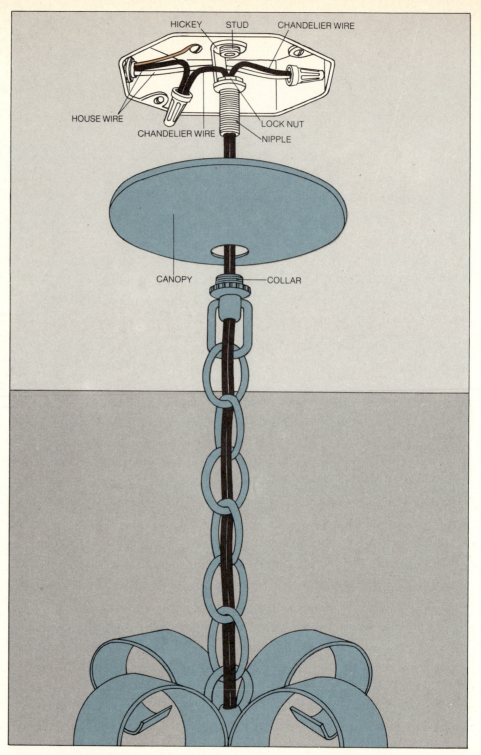

Hanging a heavy chandelier. Remove the old fixture. If the exposed ceiling box already has a stud, prepare the wiring of the box and the new chandelier by following the basic instructions on page 32 for wiring ceiling fixtures. Thread a large adapter called a hickey onto the stud in the ceiling box and attach a nipple to the hickey. Screw the nipple into the hickey far enough so that when the chandelier's collar is in place it will hold the canopy of the fixture snugly against the ceiling. Secure the nipple with a lock nut. Push the chandelier wires through the collar, canopy, nipple and the lower part of the hickey. Using wire caps, attach the chandelier wires to the house wires in the ceiling box—neutral to the white wire, hot to the black wire. Raise the canopy and secure it against the ceiling by screwing the collar onto the end of the nipple.

Rewiring a Chandelier

1 **Replacing the main cord.** Shut off the power, remove the chandelier from the ceiling box and detach its main-cord wires from the house wiring. Locate the branch wires connecting the main-cord wires to the chandelier's sockets. On some chandeliers, the connections are clearly visible on the outside of the fixture, but more often they are concealed. In the drawing at right, for example, they are hidden in a caplike recess at the lower end of the chandelier's main tube.

Open the part of the fixture where the wire connections are made and disconnect the socket wires from the main-cord wires. Pull out the old main cord. Using a length of 16-gauge stranded rip cord for a new main cord, code one of the wires as black, and snake the cord through the chandelier's center tube. Strip ½ inch of insulation from both ends of the new main-cord wires and twist the exposed strands.

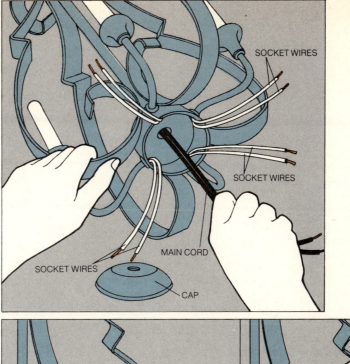

SOCKET WIRES

SOCKET WIRES

SOCKET WIRES

MAIN CORD

CAP

2 **Replacing socket wiring.** Many chandeliers have sockets, like lamp sockets, with terminal screws (*near right*). To replace wires, slip the insulating sleeves off the sockets and detach the wires. Replace them with 16-gauge stranded wire of the same length, the ends stripped ½ inch and twisted. The wire attached to the brass-colored terminal is hot; code it with black tape. If terminals are not color-coded, the wire connected to the base terminal should be marked black.

Some chandeliers have sockets with preattached wires (*far right*). This type must be unscrewed from the threaded tube in the chandelier arm and replaced. Snake the wires of a new prewired socket back through the chandelier and screw the socket to the tube. Code black the free end of the wire connected to the brass-colored terminal visible inside the base of the socket.

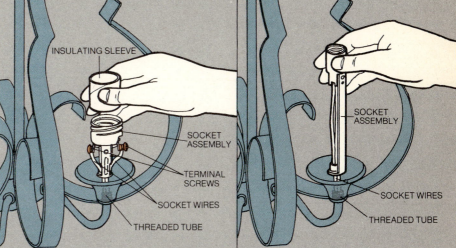

INSULATING SLEEVE

SOCKET ASSEMBLY

TERMINAL SCREWS

SOCKET WIRES

THREADED TUBE

SOCKET ASSEMBLY

SOCKET WIRES

THREADED TUBE

3 **Hooking up the new wiring.** Take the unattached ends of the socket wires you have coded as hot and twist them together with the end of the main-cord wire that you have coded with black tape. Twist together the neutral socket wires and the other main-cord wire. If you had to open a part of the chandelier to make the connections, conceal joined wire ends there and close it. Reattach the fixture to the ceiling box following the directions on the opposite page. Connect the black-coded main-cord wire to the black house wire and the other cord wire to the white house wire.

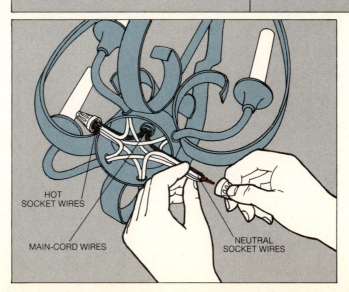

HOT SOCKET WIRES

MAIN-CORD WIRES

NEUTRAL SOCKET WIRES

Fluorescent Light Fixtures

Although the initial cost of bulbs and installation may be greater, fluorescent lighting not only gives more uniform illumination than incandescent lighting but also is so much more efficient in the use of energy that in the long run it is less expensive. A standard 40-watt fluorescent tube, for example, produces as much as six times more light than a 40-watt incandescent bulb. In normal use, the same tube will also last about five times longer than an incandescent bulb.

There are drawbacks to fluorescents, however. The color quality of the light does not exactly duplicate the "natural" effect of incandescents, although some types come very close. And design is limited by the shape of the tubes: straight, circular and, less commonly, U-shaped. Therefore, fluorescents are most often used over workbenches, counters and desks where high-intensity light is needed, in living areas to provide indirect lighting or to accent a special feature (pages 90-93), or above translucent panels in a kitchen, bath or family room to create a luminous ceiling (page 94).

A variety of fluorescent tubes. The quality of light produced by various fluorescent tubes is described in the chart at right. Those classed as cool white, warm white and daylight white produce the maximum amount of light per watt. Deluxe tubes are designed to bring out certain colors with maximum intensity, but may produce 30 per cent less light per watt than the other types. Fluorescent tubes that are designed for growing plants indoors also produce about 30 per cent less light per watt than the standard ones.

Fluorescent lights operate on a principle quite different from that of incandescents. An incandescent bulb emits light because the filament inside the bulb becomes hot and luminous as soon as the fixture is switched on. A fluorescent tube, on the other hand, works by a multistep electrical process: the switch starts an electric current in a gas, causing it to emit light waves that are not readily perceived by human eyes; these waves strike chemicals coating the inner surface of the tube, "exciting" these phosphors so that they emit visible light.

To start the process, a device called a ballast provides a momentary extra surge of voltage. In older types there is also a starter, a device that helps the ballast produce the high initial voltage; such lights may take several seconds to glow brightly. Newer fluorescent lights, called rapid-start or instant-start, produce sufficient initial voltage without a starter, and their tubes light up almost immediately. Although there are differences in wiring (page 38), all types of fluorescent lights are equally easy to hook up to your house wiring and attach to a wall or ceiling box (opposite).

A malfunctioning fluorescent fixture with either a straight or a circular tube is simple to repair because its major components—tube, tube holders and ballast—are easily removed. The cause of most problems can be diagnosed quickly by referring to the troubleshooting chart on page 39. If the cause of the problem is not obvious, you still may be able to diagnose it yourself by substituting new parts, one at a time. Before starting any work, shut off power in the circuit that serves the light. Be sure to use new parts of the same wattage as those you are replacing. First replace the tube. If that does not solve the problem and the lamp has a starter, replace the starter.

Put in a new ballast, the most costly replacement part, only as a last resort. If the ballast is faulty, you may find that it is only slightly more expensive to buy an entire new fixture. However, if you wish to install a new ballast in a straight-tube light, remove the lid from the metal box, called the channel, that holds the light to wall or ceiling; the ballast is either mounted on the inside of the lid or on the channel itself. The ballast in a circular fixture can be reached only by removing the entire fixture from the ceiling box. To avoid making a mistake in wiring a new ballast, transfer the wiring connections one at a time, checking against the diagram that is printed on the new ballast.

A Color Balance to Suit Your Purpose

Type	Color Characteristics
Cool white	Intensifies white, gray, blue and green. Does not blend well with incandescent light.
Warm white	Slightly distorts all colors but blends well with incandescent light.
Deluxe warm white	A close approximation to incandescent light. Greatly intensifies warm colors—red, orange and yellow. Blends well with incandescent light.
White	A compromise between warm white and cool white. Dulls warm colors slightly. Blends adequately with incandescent light.
Deluxe cool white	A close approximation to daylight giving a very natural color rendition. Intensifies all colors, especially cool colors. Does not blend well with incandescent light.
Daylight white	Produces a cold, blue-white light that dulls all warm colors. Blends poorly with incandescent light.
Soft white/natural	Produces a pinkish-white light that dulls all cool colors and intensifies warm colors. Blends adequately with incandescent light.
Plant-growth tubes	Generates extra amounts of colors needed for plant growth. Not intended for general lighting use.

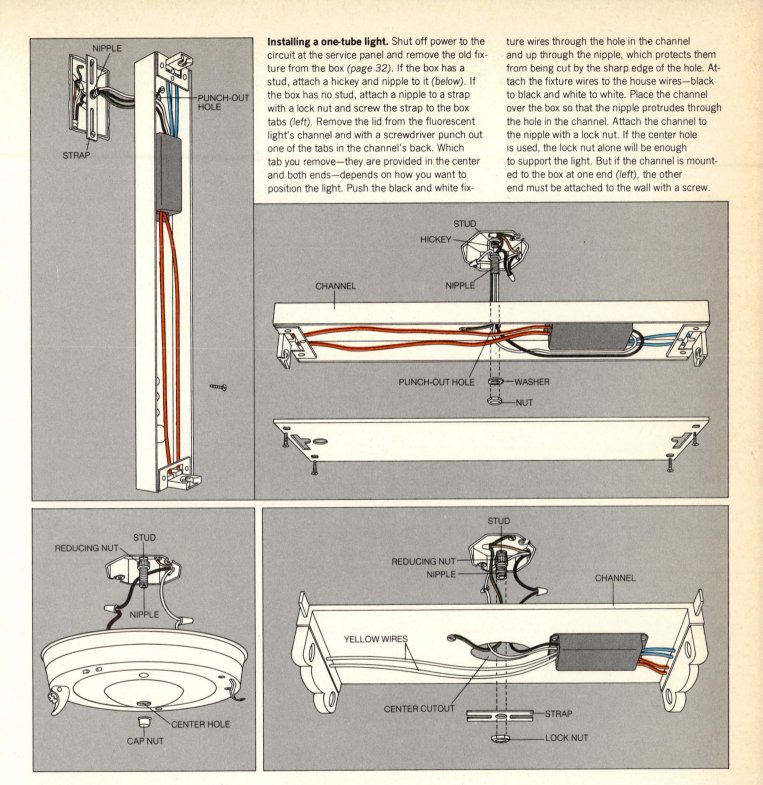

Installing a one-tube light. Shut off power to the circuit at the service panel and remove the old fixture from the box (*page 32*). If the box has a stud, attach a hickey and nipple to it (*below*). If the box has no stud, attach a nipple to a strap with a lock nut and screw the strap to the box tabs (*left*). Remove the lid from the fluorescent light's channel and with a screwdriver punch out one of the tabs in the channel's back. Which tab you remove—they are provided in the center and both ends—depends on how you want to position the light. Push the black and white fixture wires through the hole in the channel and up through the nipple, which protects them from being cut by the sharp edge of the hole. Attach the fixture wires to the house wires—black to black and white to white. Place the channel over the box so that the nipple protrudes through the hole in the channel. Attach the channel to the nipple with a lock nut. If the center hole is used, the lock nut alone will be enough to support the light. But if the channel is mounted to the box at one end (*left*), the other end must be attached to the wall with a screw.

Installing a circline fixture. Remove the old fixture. Thread a reducing nut onto the stud in the box and then screw a nipple into the reducing nut. Attach the fixture wires to the house wires —black to black and white to white—and fold them gently into the box. Place the fixture against the box so that the nipple protrudes through the center hole. Thread a cap nut onto the nipple and tighten until the fixture is firmly on the ceiling.

Installing a larger fixture. A fluorescent light with two or more tubes may have a large cutout in the center of the channel. Such a fixture can be attached only to an octagonal box that has a stud. Thread a reducing nut onto the stud and screw a nipple into it as for a center-mounted ceiling fixture (*above, left*). Pull the black and white fixture wires through the cutout in the channel and attach them to the black and white wires in the ceiling box as shown. Place the fixture over the box so that the nipple protrudes through the cutout, and thread a strap onto the nipple to hold the channel to the box. Secure the strap, which will hold the channel to the ceiling, with a lock nut.

The Three Types
of Fluorescent Lights

Rapid-start. This type, the most popular fluorescent fixture for home use, requires no starter and lights up without the few seconds' delay of the starter type. If a rapid-start malfunctions, the cause may be improper grounding. For proper operation, the tube should be no more than ½ inch from a grounded metal strip, usually the channel or a metal reflector attached to the fixture. A rapid-start tube may also malfunction if it is turned on frequently but only for brief periods at a time. Such operation interferes with key electrical mechanisms inside the lamp and it may take several seconds to light. If such a tube does light up, let it run for several hours. If it continues to work properly, the malfunction has cured itself and no replacement parts are needed.

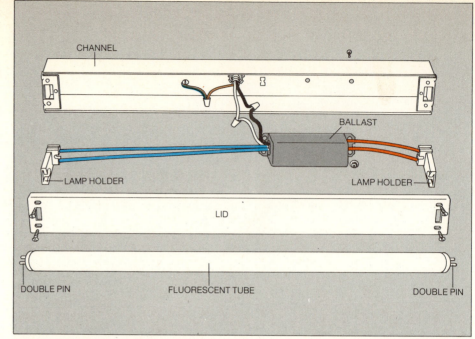

CHANNEL
BALLAST
LAMP HOLDER
LAMP HOLDER
LID
DOUBLE PIN
FLUORESCENT TUBE
DOUBLE PIN

Starter type. Many low-wattage fluorescent lights, as well as larger ones of an older type, contain starters, a small canister in a socket near one lamp holder. To replace a worn-out starter, turn off the power, remove the fluorescent tube, turn the starter counterclockwise and pull it from its socket. Be sure the new starter matches the wattage that is printed on the tube.

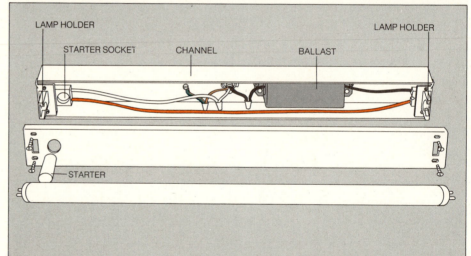

LAMP HOLDER
LAMP HOLDER
STARTER SOCKET
CHANNEL
BALLAST
STARTER

Instant-start. This type is relatively trouble-free, but when it does malfunction the trouble often arises because of the high voltage needed to start it. Even when the tube is partially burned out and blackened at one end, the high voltage will allow it to operate—producing brilliant orange flashes—but it should be replaced immediately or the ballast may be damaged. Failure to operate at all may be due simply to improper insertion of the tube in the lamp holders. As a safety measure, one holder has a spring-operated switch that shuts off current if the pin at the end of the tube is not fully inserted. To make sure the pins are seated, jiggle the tube in its holders.

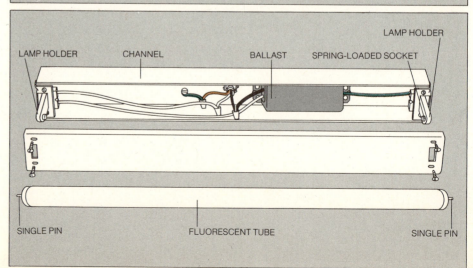

LAMP HOLDER
LAMP HOLDER
CHANNEL
BALLAST
SPRING-LOADED SOCKET
SINGLE PIN
FLUORESCENT TUBE
SINGLE PIN

Troubleshooting Fluorescent Fixtures

Problem	Possible Causes	Solution
Tube will not light	Fuse blown or circuit breaker switched off	Replace fuse or reset circuit breaker.
	Tube worn out	Replace tube.
	Dirt on tube	Remove tube and clean it with damp cloth: let dry before replacing.
	Tube pins not making proper contact with lamp holders	Rotate tube in holders for starter and rapid-start types. For instant-start type, make sure pins are fully seated in sockets.
	Incorrect tube for ballast	Check that tube wattage is same as that shown on ballast.
	Incorrectly wired ballast	See diagrams opposite or check wiring diagram on ballast. Rewire if necessary.
	Defective starter	Replace starter.
	Defective ballast	Replace ballast.
	Low voltage in circuit	Have power company check house voltage.
	Air temperature below 50°	Install low-temperature ballast.
Ends of tube glow but center does not light	Defective starter	Replace starter.
	Incorrectly wired ballast	See diagrams opposite or check wiring diagram on ballast. Rewire if necessary.
	Inadequate ground (especially in rapid-start type)	Check attachment of fixture's ground wire.
Tube flickers or blinks	Normal with new tube	Should improve with use—if not, replace starter.
	Tube pins not making proper contact with lamp holders	Rotate tube in holders for preheat and rapid-start types. For instant-start, make sure pins are fully seated in sockets.
	Tube worn out	Replace tube.
	Air temperature below 50°	Install low-temperature ballast.
Fixture hums or buzzes	Ballast wires loose or incorrectly attached	Tighten connections and check wiring against diagram on ballast.
	Incorrect ballast	Replace with ballast of correct type and wattage.
Brown or grayish bands about 2 inches from ends of tube	Normal	
Dense blackening at ends of tube	Tube worn out	Replace tube. If tube is new, replace starter instead.
Slight blackening at ends of tube	Tube nearly worn out	Replace tube.

Switching Switches

The wall switch that turns a light or appliance on and off is one of the sturdiest devices you will ever own—under normal use, a good one lasts at least 20 years. Long before a switch breaks down, however, you may decide to replace it with a newer, more sophisticated model. Some switches have special grounding terminals that are designed to prevent shocks from switch parts and cover plates. Or you might prefer one of the special-purpose switches described on page 42—a mercury switch, which is completely silent and lasts a lifetime; a switch that turns itself on and off at preset times; or a dimmer switch, which simultaneously matches the light level to your mood and saves money.

No matter what switch you choose—and even if you are only replacing one that has gone bad—the installation procedure is essentially the same. Turn off the power, disconnect the old switch and buy a new switch of the right type—the rules of thumb on this page will enable you to identify the four basic types. Check the data stamped on the mounting strap of the new switch to be sure that it matches the old in voltage and amperage ratings and meets modern safety requirements (opposite, bottom). The replacement switch may have terminals of different types or in different positions; the information at the top of the opposite page will help you choose the best switch for your needs.

If the new switch is identical with the old one, you need only hook up the same wires to the same terminals. Even if it is not, standardized color-coding of wires and terminals makes it easy to connect switches correctly.

If your house has aluminum wiring, you should have an electrician replace the switch for you (page 25). A switch that has special features designed for use with aluminum wiring must be used.

The Four Basic Switches

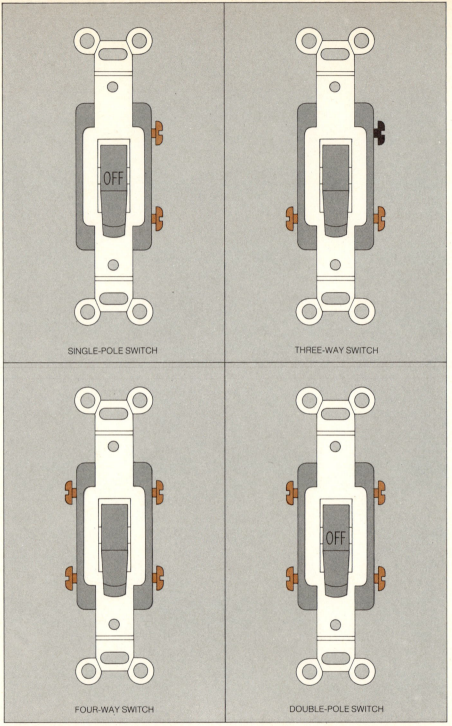

SINGLE-POLE SWITCH

THREE-WAY SWITCH

FOUR-WAY SWITCH

DOUBLE-POLE SWITCH

Telling the switches apart. A single-pole switch, the commonest of all switches, controls a light or receptacle from one location. It has two brass-colored terminals and "on" and "off" markings on the handle, or toggle. Three-way switches, used in pairs to control a light or receptacle from two locations, have three terminals: one black or copper-colored and two brass- or silver-colored. There are no "on" and "off" markings. A four-way switch works with three-way switches to control a light or receptacle from three or more locations. It has four brass-colored terminals and no "on" and "off" markings. A double-pole switch, sometimes used to control 240-volt appliances, can be mistaken for a four-way switch—both have four brass-colored terminals—but the double-pole switch has "on" and "off" markings.

Terminals for Convenience

A range of choices. Though all switches are standardized in operation and fit standard-sized outlet boxes, they vary widely in the placement and type of their terminals. The single-pole switches at right illustrate these variations. All work in exactly the same way, but you can choose the terminal type that you find easiest to wire and the terminal positions that permit you to arrange the wires conveniently in the outlet box.

Screw terminals, the commonest, are usually located on one or both sides of the switch housing, but may be on the top and bottom ends or recessed in the front. Push-in terminals, always located on the back of the housing, eliminate the need for a fastening loop since they secure the wire with spring clamps when it is inserted into the hole. A rectangular release slot is used to disconnect the wire *(page 27)*.

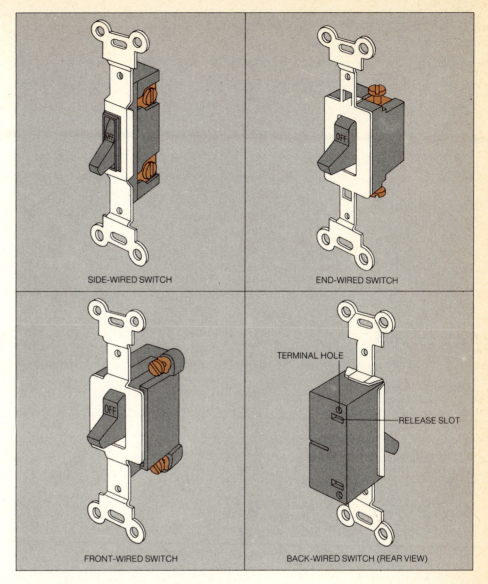

SIDE-WIRED SWITCH

END-WIRED SWITCH

FRONT-WIRED SWITCH

BACK-WIRED SWITCH (REAR VIEW)

TERMINAL HOLE

RELEASE SLOT

Reading a Switch

Facts to look for. The mounting strap of a switch is stamped with data on safety tests and operating characteristics: The Underwriters' Laboratories, Inc. listing (abbreviated UND. LAB. INC. LIST.) or, in Canada, the symbol of the Canadian Standards Association (a CSA monogram), which indicates that the switch has met standardized tests. The maximum voltage and amperage at which the switch may be used is indicated (15A-120V means the switch can control up to 15 amperes of current at voltages up to 120 volts). An AC ONLY stamp, or an AC designation in the rating (for example, 15A-120VAC), indicates the switch can be used only in the AC systems that now are almost universal. And, if the switch can safely be used with aluminum wiring, the mounting strap will have a CO-ALR stamp.

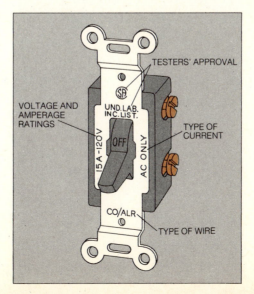

TESTERS' APPROVAL

VOLTAGE AND AMPERAGE RATINGS

UND. LAB. INC. LIST.

TYPE OF CURRENT

TYPE OF WIRE

Specialized Switches That Upgrade Your Wiring

Simply replacing a switch with one of a variety of specialized types can create dramatic lighting effects, reduce electricity bills or add convenience. Some new switches last longer, operate more quietly and incorporate safety features that were not available when older houses were wired. Others turn current off automatically when it is not needed, reduce it when lights are lowered, or give pilot-light warnings when unseen lights are left burning. Before replacing an existing switch, however, remember that the four basic types of switches shown on page 40 are not interchangeable; a single-pole switch must be replaced with another single-pole switch, never with a three- or four-way switch.

☐ QUIET SWITCH. In the AC-DC switches still found in many houses, powerful springs snap the electrical contacts together and apart very rapidly with a loud click. The rapid action was required with direct current, which has almost disappeared from U.S. and Canadian homes; the newer AC ONLY switch opens and closes more slowly and gently so it operates quietly and lasts longer.

☐ MERCURY SWITCH. A further improvement on the quiet switch, the mercury switch is completely silent, an especially desirable feature in bedrooms and nurseries; it is also almost wear-free—one common make is guaranteed for 50 years. A mercury switch can be identified by the word TOP stamped at one end of the mounting strap; the switch will not operate properly if it is installed crooked or upside down.

☐ DIMMER SWITCH. Devices to alter light levels have long been used; their modern versions not only control the intensity of an incandescent or fluorescent light, but also repay their initial expense by saving electricity when the light is dimmed. Incandescent dimmers are relatively easy to install (page 49). Fluorescent dimmers, somewhat more complex, include both a dimmer switch and a special ballast (pages 50-51).

☐ TIME-DELAY SWITCH. When this type of switch is flicked to "delay," the light it controls stays on about 45 seconds, then turns off automatically. It is particularly useful in basements, garages and patios.

☐ MANUAL TIMER SWITCH For a delay period longer than that provided by the time-delay switch (above), use a spring-wound manual timer switch, which turns off both lights and appliances. These switches come in a variety of models, with maximum delay settings ranging from several minutes up to 12 hours; in addition, many switches have "hold" positions that keep them on indefinitely. These switches are often used to control kitchen or bathroom ventilation fans.

☐ TIME-CLOCK SWITCH. Completely automatic "on" and "off" control—to turn an air conditioner on before a room is occupied, or to turn lights on and off at preset times to discourage burglars when a house is vacant—has made this type popular. Its clock motor requires an extra neutral wire connection (page 45).

☐ LOCKING SWITCH. When, for security or safety reasons, you need to limit access to light switches or switch-controlled power tools, you can replace any of the four basic switch types with a locking switch. These switches can be turned on and off only by inserting a special key (which comes with the switch) into a slot on the face of the switch. On a locking switch, approval by Underwriters' Laboratories means more than acceptable safety in the normal sense; to pass the UL tests, a locking switch must be tamperproof and designed so that a metal object inserted into the key slot cannot touch electrical parts.

☐ LIGHTED-HANDLE SWITCH. A tiny built-in neon bulb that glows when the handle is turned to "off" makes this switch easy to find in a darkened room or stairway. The indicator light lasts for many years, uses hardly any electricity and needs no additional wiring.

☐ PILOT-LIGHT SWITCH. Unlike a lighted-handle switch, which glows when it is set to "off," a pilot-light switch has a bulb that shines when the switch is turned to "on." Available in single-pole and three-way versions, it is especially useful for controlling basement, attic and outdoor lights, as well as appliances that cannot be seen from the switch location. Some models have a separate pilot light, others have the light in the handle; for wiring instructions, see page 45.

The Single-Pole Switches

Single-pole switches are the basic type most frequently used in the home—and therefore, the ones most frequently replaced, either because a switch has failed or because you want to substitute a newer type. Before deciding that a switch has failed, however, check the light bulb it controls and the fuse or circuit breaker that controls it.

Most single-pole switches are replaced as shown below and on the following page—a job requiring one screwdriver, pliers and two inexpensive testers. The replacement may be fitted with a special grounding terminal, a protection against shock that most older switches do not have, and this extra connection requires an additional step in installation (page 44). And some special devices such as time-clock and pilot-light switches require a neutral wire connection and must be installed as shown on page 45.

When you begin the replacement of any switch, turn off power to the switch, unscrew the cover plate and loosen the two screws holding the switch in the box until you can pull the switch out. The screws should not come out of the mounting strap (fiber washers hold them in the slots). Grip the ends of the mounting strap and pull the switch out of the box until the wires are fully extended.

2 **Removing the switch.** To be sure the current is indeed shut off, touch one probe of a voltage tester to the metal shell of the outlet box and the other probe to each of the brass terminals in succession. If the switch has push-in terminals, insert the probe into the release slots. The tester should not glow on either terminal. If it does, you have not disconnected the fuse or turned off the circuit breaker that controls the circuit; return to the service panel and find the right one. When you are certain that the power is off, loosen the terminals with a screwdriver and remove the wires, using long-nose pliers to open the loop on the end of each wire. (The method for disconnecting push-in terminals is described on page 27.)

Replacing a Switch

1 **Identifying the wiring.** With the wires exposed, you will find one of the two wiring variations that determine how the new switch is to be hooked up. In one variation, called middle-of-the-run wiring (below, left), at least two cables enter the box. Two black wires, or one black and one red wire, are the hot wires attached to the switch terminals; you will also see white wires and generally bare copper wires connected with wire caps. The white wires are neutral, the bare ones are ground wires; neither should be disconnected or pulled out of position for this job. In the second variation, called a switch loop (right), only one cable enters the box, and one black and one white wire are connected to the switch. Here, the white wire is not neutral; it is hot and should be marked as such with black electrical tape or black paint near the end of the insulation. If the wire has not been so identified, code it before replacing the wires.

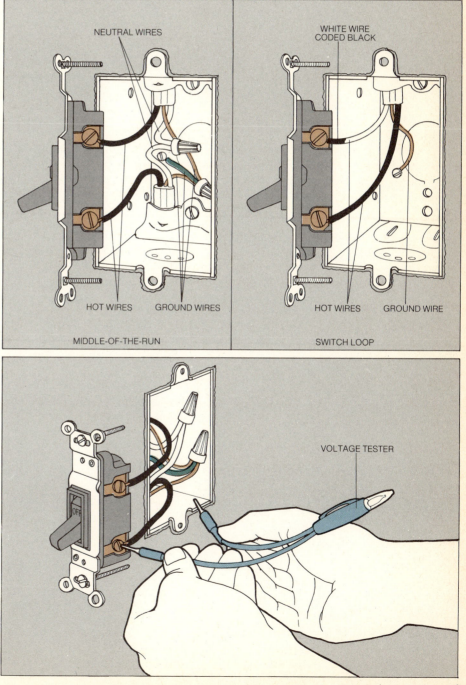

NEUTRAL WIRES

HOT WIRES   GROUND WIRES

MIDDLE-OF-THE-RUN

WHITE WIRE CODED BLACK

HOT WIRES   GROUND WIRE

SWITCH LOOP

VOLTAGE TESTER

3 **Testing the switch.** If you are replacing the switch because you think it faulty, use a continuity tester for two checks on the internal wiring of the switch. First, apply the alligator clip and probe of the tester to the switch terminals. Move the switch handle back and forth between the "off" and "on" positions. On a good switch, the tester will light at "on," but not at "off." Second, fasten the alligator clip of the tester to the metal mounting strap of the switch and touch the probe to each of the switch terminals in succession, moving the handle from the "off" to the "on" position at each terminal. On a good switch, the tester will not light in any position. If the switch fails either of these tests, it must be replaced.

4 **Installing the new switch.** Align the replacement switch vertically. A single-pole switch should be off when the handle is down and on when it is up. Connect the two hot wires to the terminals, either wire to either terminal. Push the switch back into the outlet box, folding the slack wire behind the switch, and fasten it with the mounting screws.

If the box is slightly tilted—as it generally is—change the position of the screws in the wide mounting slots to get the switch straight. If the box is recessed in the wall, circular tabs called plaster ears, located at the corners of the mounting strap, will keep the switch handle flush with the wall. If the box is flush and the ears get in the way of the cover plate, take them off with pliers.

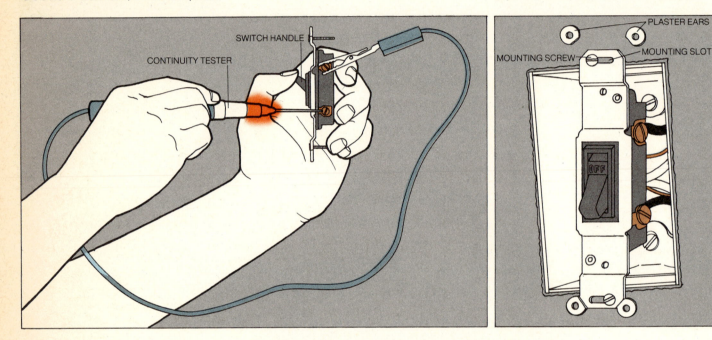

SWITCH HANDLE

CONTINUITY TESTER

PLASTER EARS

MOUNTING SCREW — MOUNTING SLOT

Grounding Switches

A safer grounding scheme. In the past, the metal parts of switches were grounded only by mounting screws that fastened them to grounded outlet boxes. Switches are now available with a more reliable ground connection: a separate grounding screw terminal, which is identified by green color-coding on the screwhead or by the letters GR next to the screw hole. If your dealer has such switches in stock, they should be used as replacements.

To ground one of these switches in an outlet box wired with plastic-sheathed cable, run a short length of either bare wire or insulated green wire from the green grounding terminal to the wire cap that links the bare ground wires of the cables and the box. If armored cable serves the box, you may find no separate ground wire. In that case, run the wire from the green terminal directly to a grounding screw in the box.

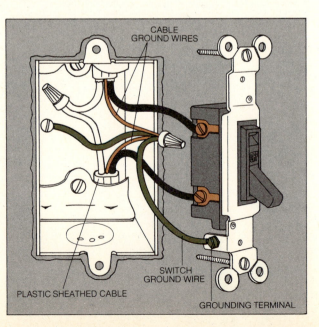

CABLE GROUND WIRES

PLASTIC SHEATHED CABLE

SWITCH GROUND WIRE

GROUNDING TERMINAL

Time-Clock Switches

1 **Preparing the box wires.** A wall-mounted time-clock switch can replace a single-pole switch only in middle-of-the-run wiring (*page 43, Step 1*), where a neutral wire is available for the clock motor. If this requirement is met, remove the existing switch. Observing the precautions described on page 22, turn the power back on. Use a voltage tester to determine which of the two black wires in the outlet box is now hot; this is the incoming wire from the service panel. Turn the power off. Mark the incoming wire with tape and remove the wire cap from the neutral wires.

2 **Connecting the switch.** Attach the special mounting plate to the box. Straighten out the wire loops at the ends of the black cable wires. Using wire caps, connect the incoming black wire to the black wire of the switch, the outgoing black wire to the red switch wire, and the two white cable wires to the white switch wire. Fasten the switch to the mounting plate, restore power and follow the manufacturer's directions to test the switch.

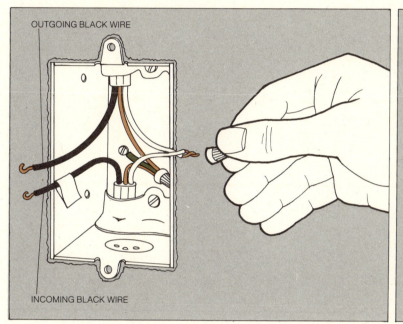

OUTGOING BLACK WIRE

INCOMING BLACK WIRE

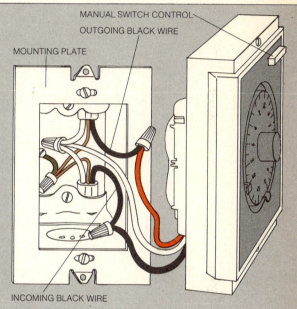

MANUAL SWITCH CONTROL
OUTGOING BLACK WIRE
MOUNTING PLATE
INCOMING BLACK WIRE

Pilot-Light Switches

Connecting the switch. Like time-clock switches, pilot-light switches can replace ordinary single-pole switches only in middle-of-the-run wiring where a neutral wire is available, in this case for the pilot-light bulb. Remove the old switch and prepare the box wiring as you would for a time-clock switch (*Step 1, above*). Most pilot-light switches have three brass terminals and one silver one, with two of the brass terminals joined by an exposed brass strip. Connect the outgoing black cable wire to either of the joined terminals and the incoming black wire to the brass terminal on the opposite side. Connect the silver terminal to the white cable wires with a white jumper.

Some pilot-light switches have a bulb in the handle, and two rather than three brass terminals. Connect each brass terminal to one black cable wire. If the pilot light stays on when the switch is off, reverse the black-wire connections.

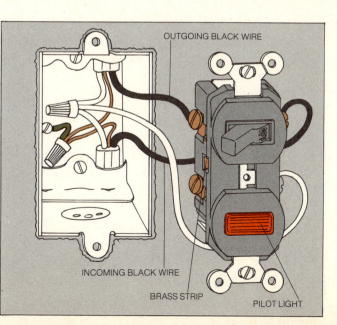

OUTGOING BLACK WIRE

INCOMING BLACK WIRE

BRASS STRIP

PILOT LIGHT

Replacing a Three-Way or Four-Way Switch

If you are replacing a three-way switch because it is defective, you must first determine which of the two switches is the faulty one. Check three-way switches with a continuity tester as shown in the drawings at bottom.

However, when a four-way switch is used in conjunction with a pair of three-way switches (shown opposite), you will have to remove all of the cover plates to determine which of the devices are three-way switches and which is the four-way. A three-way switch has three terminals: a dark-colored one, the common terminal, and two lighter-colored terminals for traveler wires. (On back-wired three-way switches the common terminal is marked COM. or COMMON.) A four-way switch has four brass terminals for traveler wires connecting it to three-way switches. If both three-way switches check out satisfactorily, then you will know that the four-way switch is defective and should be replaced.

Before you put a new switch in, pay close attention to the way the wires are connected to the old switch. Like single-pole switches, the three- and four-way switches interrupt only the hot wire. You may find black, red and white wires connected to either type of switch, but all of them are hot wires regardless of color. White wires that are connected to either switch should be coded black as shown in the drawing at top and insets opposite. Other wires passing through the box go to other devices and should not be disconnected when replacing the switch.

Quiet-action and mercury types are available as replacements for both three- and four-way switches, and there are also three-way switches equipped with lighted handles, dimmer controls (page 49) and pilot lights (pages 42, 45).

Replacing a three-way switch. Shut off the power to the circuit and without disconnecting the wires remove the switch from its box. Double-check that power is off by touching one probe of a voltage tester (page 43) to each of the three switch terminals in turn, while the other probe is touching the grounded box. Before disconnecting the wires, use a piece of masking tape to designate the wire that is attached to the common terminal—it is usually black or copper-colored, but is always darker than the brass- or silver-colored traveler terminals. Disconnect all three wires. Then connect the wires to the new switch, attaching the marked wire to the dark-colored common terminal. Either traveler wire can be connected to either traveler terminal.

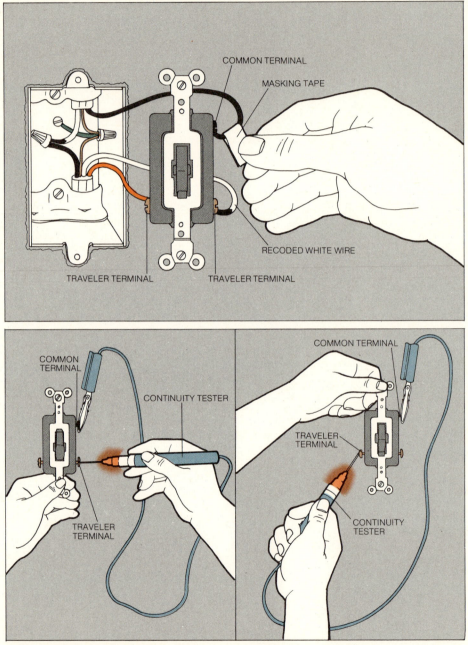

Testing three-way switches. To find which of a pair of three-way switches is faulty, shut off the power, disconnect one switch (drawing, top), and attach the clip of a continuity tester (page 44) to the common terminal. Place the tester probe on one traveler terminal (above, left) and move the switch toggle up and down. The tester should light when the toggle is in only one position, either up or down. Leaving the toggle in the position that showed continuity, touch the probe to the other traveler terminal (above). The tester should not light in this position, but should light when the toggle is flipped to the opposite position. If the switch passes both these tests it is good; reconnect it, then remove and check the other switch in the same manner.

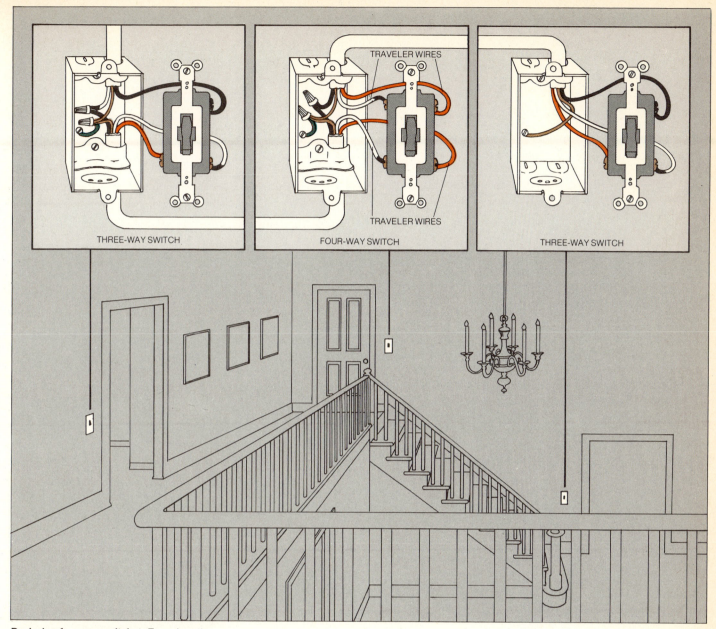

THREE-WAY SWITCH

FOUR-WAY SWITCH

TRAVELER WIRES

TRAVELER WIRES

THREE-WAY SWITCH

Replacing four-way switches. To replace the four-way switch used with two three-way switches, shut off power and remove the cover plates from all three switches to make sure which is the four-way device. The hot traveler wires to the four-way switch may be black or red, or white re-coded black. Disconnect the traveler wires from the terminals at the top of the switch and transfer them to the terminals at the top of the replacement switch. (Either traveler wire can be connected to either terminal.) Then repeat the procedure for the wires on the bottom terminals.

If the replacement switch does not work properly when wired as shown in the center inset above, it may have a switching mechanism requiring a different wiring configuration. In that case, shut off power, loosen the two wires on either side of the switch and reverse the connections.

Decorating with Light

Nowhere is lighting more critical than in the theater: It is used to underscore a play's mood, draw attention to certain aspects of costume or setting, and even reveal the personalities of the characters. Yet there is nothing unique in theatrical lighting. The evocative powers of light can be just as pronounced in a home as on a stage—and more and more homeowners are exploiting its potential for drama.

Although beautiful lamps and crystal chandeliers still have their place in modern lighting, the fixtures themselves often are no longer the focal points of a room. Instead, the emphasis today tends to be not on the light source but on where the light is directed and what it accomplishes.

Putting light to work in this way is developed in two stages. First "key lights" are positioned to highlight certain objects or limited areas. Then "fill light" is provided—general illumination that covers the area evenly. Both fill light and key light can be produced with almost any kind of fixture, from a table lamp to a recessed strip of fluorescent tubes; the choice depends on the effect to be achieved, the nature of the room, and your personal taste.

For smooth, even fill light, an extended source is necessary—valance lights, a strip of flood lamps or, to fill an entire room with shadowless illumination, a luminous ceiling. In some cases, this extended source is the ceiling itself, reflecting illumination directed at it from an "uplight" fixture or two.

Key lighting can be provided by a portion of an extended source—a soffit light, for example. Shaded lamps and sconces will also serve, but for the most demanding situations, spotlights are the usual choice.

The emotional impact of a lighting scheme depends on the balance between fill and key light. If fill light dominates, the effect is likely to be calm and restful; if the emphasis is on key light, more dramatic and exciting effects are achieved. A room's mood can be varied, of course, by turning certain lights on or off, or by installing dimmer switches to raise or lower the level of illumination.

In fact, the intensity of the light can have as much effect on the mood of a room as the type of light. During a large party, hosts might choose to have the maximum illumination possible. Abundant light makes a room cheerful, colors brilliant, and jewelry sparkling. But for a smaller gathering, they might choose a lower level of illumination for a sense of intimacy and relaxation.

Lights may also be used to show off a personal interest, focusing attention on an antique, a painting or an exotic plant. In addition, lights can even be used to emphasize architectural features or to make up for the lack of them. Light can bring out the pattern of a textured wall or stone pilasters, for example; or, in an open-plan apartment, it can help create the illusion of distinct rooms.

Outside the house, lighting can dramatize the best features of the landscaping or the architecture. And light that floods out of the windows of a house at night creates a feeling of warmth and welcome for visitors arriving out of the dark.

Incandescent Dimmer Switches

Dimmers provide a convenient way to adjust lighting levels to suit a particular activity or mood. They also save electricity and make light bulbs last longer (a dimmed filament operates at a lower temperature, slowing the burn-out process). Dimmers can be used to control only lights, not appliances or the receptacles into which appliances or motor-driven tools might be plugged. A dimmer switch also must be matched to the type of lighting it will control; incandescent dimmers cannot be used on fluorescent lights, which require their own devices *(pages 50-51)*. Finally, the total wattage of lights connected to the dimmer must not exceed the capacity listed on the front of each switch.

Wall-mounted dimmers like the ones shown here are designed for use only with permanent light fixtures and are available in both single-pole and three-way types. (Other types are available for use with table and floor lamps; check your electrical supply store for sockets with built-in dimmers and for dimmers that attach to lamp cords or plug into receptacles.) Three-way dimmers are designed so that only one of the two three-way switches *(page 46)* is replaced. Thus, a light on such a three-way switch circuit can be turned on and off from two locations but it can be dimmed at only one of those points.

Most wall-mounted dimmers are full-range controls that provide completely variable light settings, from a faint glow to full brightness. These switches usually have a 600-watt capacity—more than enough for normal household lighting. When a full-range light control is not needed, a less expensive high/low dimmer can be used. These switches are usually controlled by toggles rather than rotary knobs and provide only two levels of illumination: full brightness and about 30 per cent brightness. High/low switches generally have a 300-watt capacity.

The electronic components of dimmers sometimes cause interference on television sets and AM radios. Most dimmers have a built-in filtering device. If the problem persists in spite of the filter, move the radio or TV as far away from the dimmer as practical or, if possible, plug it into another circuit. If no other circuit is available, a power-line filter will have to be used. These devices, available at radio and television supply stores, are attached between the appliance cord and the receptacle to trap interference coming through the house wires.

Wall-mounted dimmer switches. Single-pole dimmer switches *(top right)* are installed in the same manner as regular single-pole switches *(page 43)*. The only difference you may find on some is the presence of wire leads instead of screw terminals; fasten the leads to the cable wires with wire caps. Most dimmers have rotary control knobs that are pushed onto the knurled shaft after the cover plate has been installed.

If you want to provide dimming control on a three-way switch circuit, use only one dimming switch; the other switch in the circuit remains unchanged. Decide which three-way switch you wish to replace. Shut off power to the circuit and remove the switch *(page 46)*. Be sure to tag the wire on the common terminal with tape *(right)*. Connect this wire to the black switch wire and the red switch wires to the traveler wires.

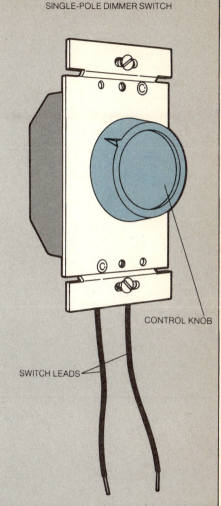

SINGLE-POLE DIMMER SWITCH

CONTROL KNOB

SWITCH LEADS

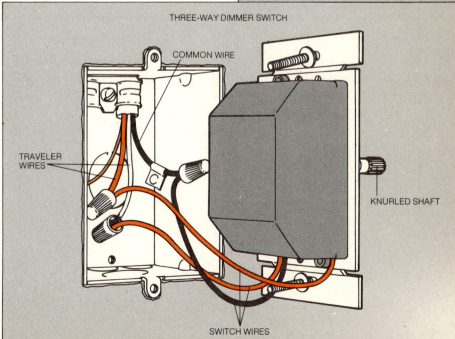

THREE-WAY DIMMER SWITCH

COMMON WIRE

TRAVELER WIRES

KNURLED SHAFT

SWITCH WIRES

Installing Dimmers for Fluorescent Lights

The bright, even and inexpensive light from fluorescent fixtures is usually exactly what you want over a workbench or kitchen counter. But in living areas, the unvarying brightness of these lights may be too much of a good thing. Substituting a fluorescent dimmer switch for the conventional single-pole one provides you with a complete range of control from a glimmer to full illumination.

A special fluorescent dimmer switch can control from one to eight fluorescent lamps at the same time. The ballast on each fluorescent fixture, however, must be replaced by a special dimming ballast. These ballasts are made only for 40-watt, 48-inch fluorescent fixtures. This is the size that is commonly used for installations such as a luminous suspended ceiling (page 94) or for valance lighting (page 92), where these dimming controls are most useful.

Each fixture must be grounded and fitted with 40-watt rapid-start lamps. An existing 40-watt rapid-start fixture like the one shown here is the easiest to convert because the lamp and lamp holders need not be replaced and the fixture is already grounded. To convert a 40-watt preheat or instant-start fixture (page 38), you must first ground the fixture and then replace the lamp and lamp holders.

The instructions that follow are for the dimming system known as a two-wire dimmer, the simplest to install; other types require additional house wiring.

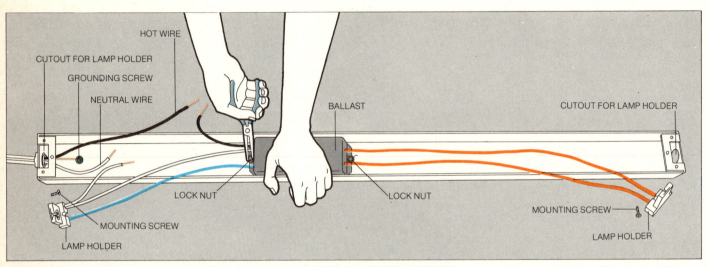

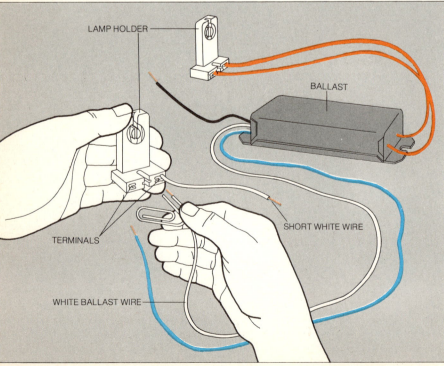

1 Removing the old ballast. After turning off the power, remove the fluorescent tube and the metal cover. Unscrew the two lamp holders and push them free. Loosen the wire caps fastening the incoming hot and neutral wires to the ballast and lamp holders; leave the ground wire in place. Remove the lock nuts securing the ballast and carefully lift out the ballast and lamp holders.

2 Rewiring the lamp holders. Remove the blue, white and red wires of the old ballast from the lamp holders. In addition to the ballast wires, one of the lamp holders has a short white wire connected to the same terminal as the white ballast wire. Leave this short wire in place when removing the other wires. If there are push-in terminals instead of screws, insert a straight, stiff length of wire such as the tip of a paper clip in the terminal alongside each wire to be removed. This will release a spring clamp so that you can pull both the wire and the paper clip out together.

Attach the white wire from the dimming ballast to the terminal that has the short white wire and connect the blue wire to the other terminal on that lamp holder. Connect the two red wires from the dimming ballast to the other lamp holder.

3 **Installing the dimming ballast.** Place the new ballast in the channel, wrap the short bare-metal ground wire from the ballast around one of the lock nuts and tighten the nuts. Screw the two lamp holders into place in the cutouts at the ends. With wire caps, connect the black and white wires to the house wiring. Make sure that the bare or green-insulated ground wire is securely attached to the channel with a screw or by fastening it under a ballast lock nut.
When dimming several fixtures, install dimming ballasts in the other channels as well. Assemble the channels as shown on page 91, using No. 14AF wire to make interconnections (page 92).

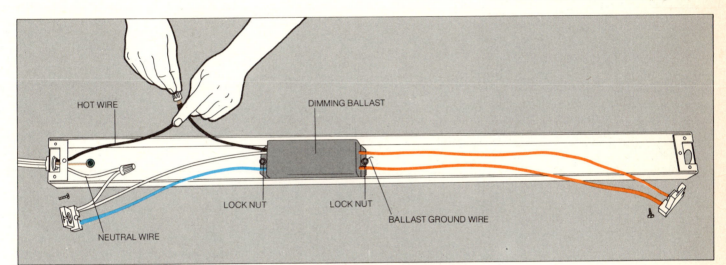

HOT WIRE

DIMMING BALLAST

LOCK NUT

LOCK NUT

BALLAST GROUND WIRE

NEUTRAL WIRE

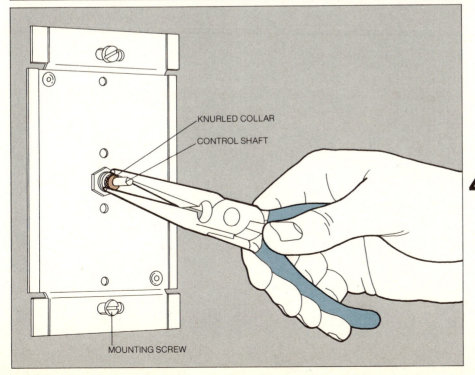

KNURLED COLLAR

CONTROL SHAFT

MOUNTING SCREW

4 **Installing the dimmer switch.** With the power to the circuit still off, replace the wall switch with the dimmer (the wiring is the same as for a regular single-pole switch, pages 43-44). Before putting on the cover plate, restore circuit power and set the minimum illumination level as follows. Rotate the control shaft to its highest setting and turn the adjustment collar fully clockwise. Rotate the control shaft to its lowest setting without turning the light off. Carefully turn the adjustment collar counterclockwise until the light begins to flicker; then turn the collar slightly clockwise until the flickering stops. This adjustment ensures that the light will not go out until the control shaft is turned completely to the off position.

Substituting New Plugs and Receptacles for Old

Plug-in receptacles and the plugs that fit into them are easy to install, whether to replace a defective one or to provide extra usefulness. A great variety of specialized receptacles are available to do the job of an ordinary one while adding safety or convenience. There are devices that combine a receptacle with a light fixture or a switch *(page 56),* nestle a receptacle below a floor, hide the cord of a clock, prevent children from poking into the holes or lock the prongs of a plug so it cannot be pulled out accidentally.

To connect a new receptacle, turn off the power and follow the wiring patterns on pages 54-55. If the replacement is identical with the old one, you will probably hook the same wires to the same terminals—but be sure the old one was wired correctly. Some old installations may be improperly grounded or not grounded at all. More often, the replacement will differ slightly from the old one.

When buying a replacement of the same basic type, make sure it matches the existing receptacle in voltage and amperage ratings *(opposite, bottom).* Many types now come with easy-to-use push-in terminals *(opposite).* For 120-volt circuits, always use a three-prong grounded receptacle even if the old receptacle was two-prong, unless the receptacle cannot be grounded. For 120/240-volt circuits, United States codes under most circumstances allow the neutral white wire to be used for grounding the appliance, but Canada's codes require a fourth wire for grounding, with a fourth prong and slot.

Plugs come in two types. Molded plugs have the cord permanently attached; terminal types are attached to a separate cord. If a molded plug breaks, replace the plug-and-cord unit or cut the plug off and install a terminal plug; do not splice a cord to a molded plug.

120-volt, grounded. The modern 15-ampere receptacle, as well as the 20-ampere receptacle for heavy-duty kitchen circuits, has an inverted U-shaped grounding slot for a matching grounding prong on the plug. The 15-ampere plug shown here will fit both 15- and 20-ampere receptacles. A 20-ampere plug has one prong angled so it fits only a 20-ampere receptacle.

120/240-volt, 30-ampere. Designed especially for clothes driers, this large receptacle supplies 240 volts for the heating coils and 120 volts for accessories such as the timer and the pilot light. In Canada, this circuit requires a fourth grounding wire, fourth prong and fourth slot.

Types of Slots and Prongs

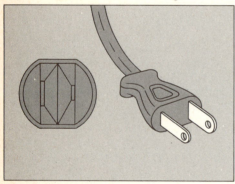

120-volt, 15-ampere, ungrounded. This receptacle should be used only for lamps and other small appliances. In many receptacles, one slot is longer than the other. When the receptacle is correctly wired, the shorter slot is hot, the longer slot neutral. All 120-volt plugs fit such a receptacle. But some appliances, such as certain TV sets, have plugs with similarly polarized prongs. These plugs can be inserted only one way, since the appliance circuits require a foolproof match of hot and neutral wires. To install the receptacle and plug, connect black wires to the terminals for the short slot and narrow prong, and white wires to the long slot and wide prong.

15 AMPERE

20 AMPERE

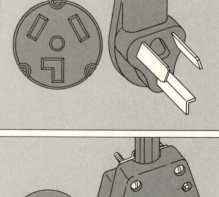

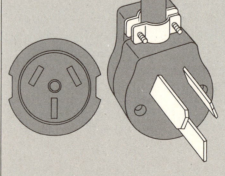

120/240-volt, 50-ampere. Electric ranges require the combination of voltages provided by this receptacle. High oven and burner settings work on the 240-volt circuit; the timer, pilot lights and built-in receptacles run on 120 volts. Canadian codes require a four-wire plug and receptacle to accommodate a ground wire.

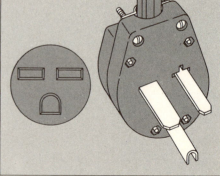

240-volt, 30-ampere, grounded. This receptacle, which supplies only 240 volts, is used mainly for hot-water heaters and 240-volt air conditioners. The plug in the drawing above is a type that may be purchased separately if a replacement is needed. It is attached to the appliance cord by the method described on page 59.

The Two Kinds of Terminals

Side-wired receptacles. Two brass-colored terminal screws for black or red wires and two silver-colored terminal screws for white wires are located on the sides of this type. A green screw for the ground wire is at the bottom.

Back-wired receptacle. In this type, wires are attached by inserting them into holes at the back of the receptacle and detached by pressing the tip of a screwdriver into a release slot *(page 27)*. A green screw terminal for the ground wire is at the bottom. Some back-wired receptacles also have screw terminals like those at left.

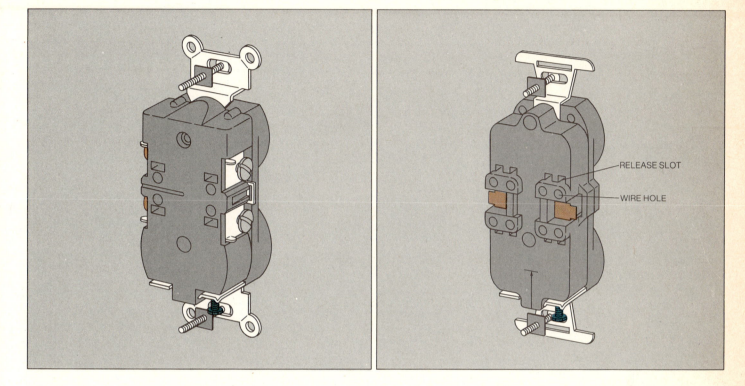

RELEASE SLOT

WIRE HOLE

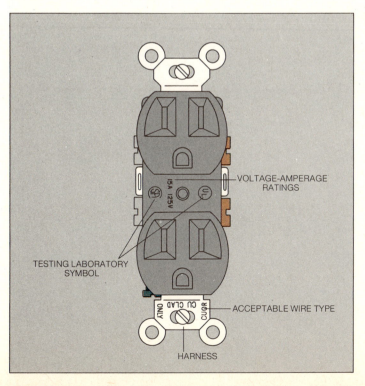

VOLTAGE-AMPERAGE RATINGS

15A 125V

TESTING LABORATORY SYMBOL

ACCEPTABLE WIRE TYPE

CU OR CU CLAD ONLY

HARNESS

Reading a Receptacle

Facts to look for. The receptacle is stamped with data on safety tests and operating characteristics: the UL monogram of the Underwriters' Laboratories, Inc., or CSA of the Canadian Standards Association, which indicates that the receptacle has met standardized tests; the maximum voltage and amperage at which the receptacle can be used; and an abbreviation indicating the metal of the wires that can be connected to it. The abbreviation CU OR CU CLAD ONLY, indicates that either copper or copper-clad wire may be used. If solid aluminum wire is acceptable, the abbreviation reads CO-ALR, CU-AL or AL-CU. In this example, 15A/125V means the receptacle can carry up to 15 amperes of current at voltages up to 125 volts, and CU OR CU CLAD ONLY indicates that uncoated aluminum wire is not to be attached.

Replacing 120-Volt Receptacles

Middle-of-the-run with plastic cable. In middle-of-the-run wiring two cables enter the box, each containing a black, a white and a bare copper wire. Connect each of the black cable wires to a brass terminal of the new receptacle; connect each white cable wire to a silver receptacle terminal. Attach one 4-inch green jumper wire to the back of the box and another to the green receptacle terminal. Connect both jumpers to the bare cable wires with a wire cap.

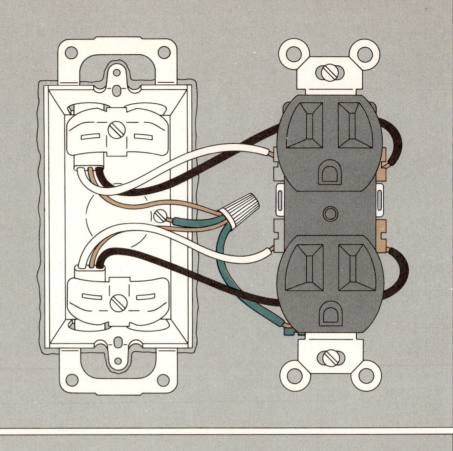

Middle-of-the-run with armored cable. Connect each black cable wire to a brass receptacle terminal and each white cable wire to a silver terminal. Attach a 5-inch green jumper wire to the back of the box with a machine screw and connect this jumper to the green terminal.

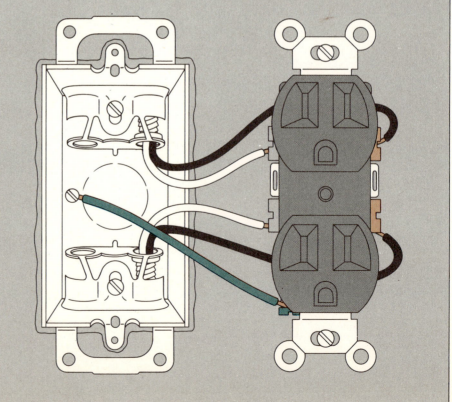

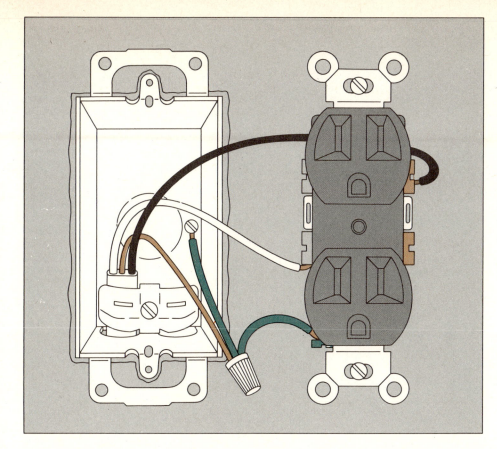

End-of-the-run wiring. In end-of-the-run wiring a single cable enters the box. If the cable is plastic-sheathed, it contains a black, a white and a bare copper wire. Connect the black cable wire to a brass terminal on the new receptacle and the white cable wire to a silver terminal. Attach one 4-inch green jumper wire to the back of the box with a machine screw and another jumper to the green receptacle terminal. Then join both jumpers to the bare cable wire with a wire cap.

Armored cable has no bare wire inside the box. To ground the installation, use a 5-inch green jumper wire and attach it to the back of the box with a machine screw. Then connect the jumper wire to the green screw on the receptacle.

From Switch to Switch-Receptacle Combination

Installing the switch-receptacle. This installation, which combines a single-pole switch with a receptacle that is always hot, is possible only if the switch to be replaced is middle-of-the-run (*opposite*), because a white neutral wire must be available for the receptacle. Connect the incoming black cable wire to one of the pair of brass terminals linked by a metal tab. Connect the outgoing black wire to the brass terminal on the opposite side of the switch-receptacle. Attach a 4-inch white jumper wire to the silver terminal and connect it to the two white cable wires with a wire cap. Attach one 4-inch green jumper wire to the back of the box with a machine screw and another jumper wire to the green terminal of the switch-receptacle. Connect both jumpers to the bare cable wires with a wire cap.

OUTGOING CABLE

INCOMING CABLE

From Light Fixture to Light-Receptacle Combination

An end-of-the-run installation. If a single cable enters the outlet box, the switch that controlled the old light fixture will control both the light and the receptacle of the new combination fixture, and the receptacle will not work unless the light is on. Connect the two black fixture wires to the black cable wire, and the two white fixture wires to the white cable wire. Attach a 4-inch green jumper wire to the back of the box with a machine screw, and connect it to the green fixture wire and the bare cable wire. Note: Because of the requirement for ground-fault interrupters in bathrooms, this type of receptacle is not allowed there.

A middle-of-the-run installation. With two cables entering the box, the combination fixture can be wired so that the switch controls only the light, while the receptacle remains hot at all times. Connect the black wire from the receptacle part of the fixture to the two black cable wires. The white wire going to the switch should be identified as hot by a dab of black paint or a piece of black tape. Connect this recoded wire to the black wire from the light. Connect the two white fixture wires to the incoming white cable wire. Attach a 4-inch green jumper wire to the back of the box with a machine screw and, using a wire cap, connect the jumper to the bare cable wires and the green wire from the receptacle. Note: Because of the requirement for ground-fault interrupters in bathrooms, this type of receptacle is not allowed there.

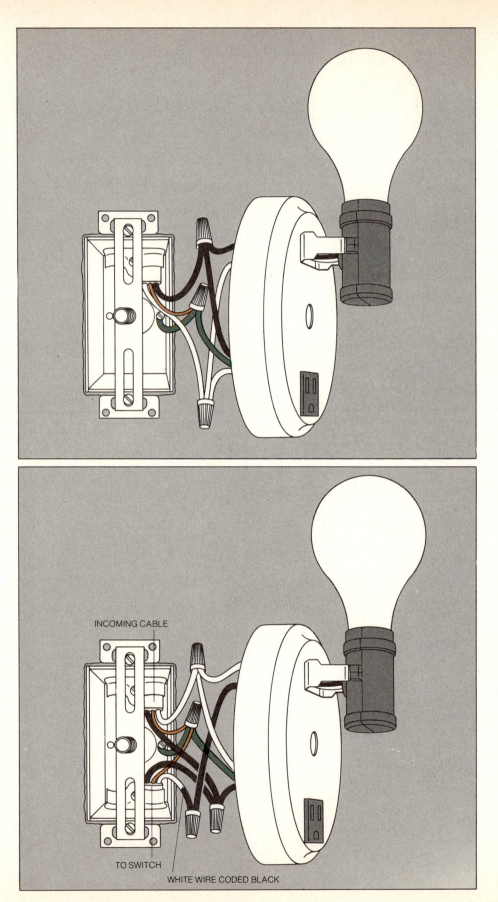

INCOMING CABLE

TO SWITCH

WHITE WIRE CODED BLACK

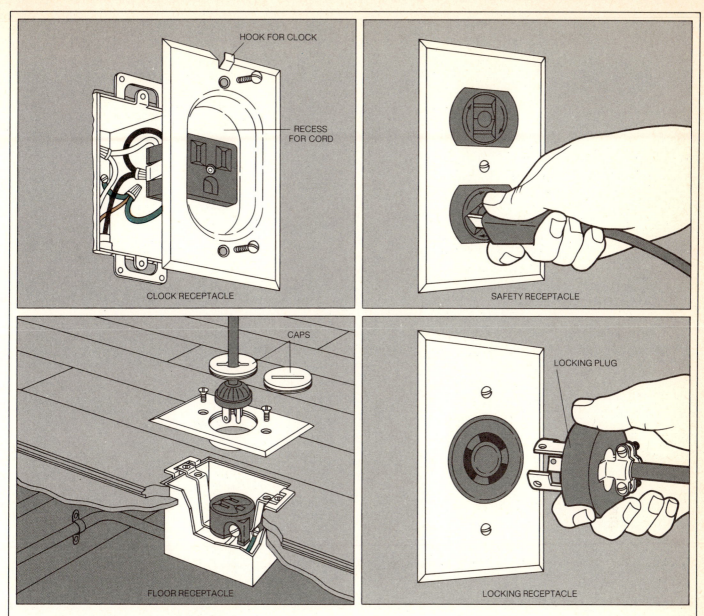

HOOK FOR CLOCK

RECESS FOR CORD

CLOCK RECEPTACLE

SAFETY RECEPTACLE

CAPS

FLOOR RECEPTACLE

LOCKING PLUG

LOCKING RECEPTACLE

Four Specialized Receptacles

A standard 120-volt receptacle can easily be replaced with a model especially designed for greater convenience or safety. The examples described on this page fit into standard-sized receptacle boxes and are wired exactly like an ordinary 120-volt grounded receptacle.

☐ CLOCK RECEPTACLE. A recessed face plate provides room for the cord of an electric clock; a built-in hook at the top of the plate makes it easy to hang the clock flush against the wall. Because the recessed plate reduces the wiring space available in the outlet box, use 3-inch rather than 4-inch jumper wires in

any installation that calls for a jumper.

☐ SAFETY RECEPTACLE. A plug can be inserted only by rotating a solid cover over the slots. When the plug is withdrawn the cover snaps back into place, making the receptacle safe from the inquiring fingers and tongues of children and pets.

☐ FLOOR RECEPTACLE. A special harness holds the receptacle well below the level of the floor. A cap fits flush with the floor and functions as a dust protector. The standard cap can be removed temporarily for an appliance such as a vacuum cleaner. For perma-

nently plugged-in appliances, such as floor lamps, the receptacle comes with an alternate cap that is attached to an appliance cord above the plug.

☐ LOCKING RECEPTACLE. Especially useful with electric drills, circular saws and other tools that are moved frequently while in use, this receptacle has a locking device that grips the prongs of a plug to prevent it from being pulled out accidentally. A corresponding locking plug, which is fitted with a clamp to secure the cord, must be substituted for the conventional plug that comes with the tool.

Connections for 240-Volt Receptacles and Plugs

A 120/240-volt, 30-ampere receptacle. Connect the white cable wire to the terminal stamped "white" by pushing the wire into the terminal and then tightening the screw. Connect the red and black cable wires to the other terminals the same way, then attach the bare cable wire to the back of the box with a machine screw.

Unlike most receptacles, this one does not have a grounding slot and terminal; its three terminals are needed for three conductors—normally hot black and red wires and a white neutral wire. But the outlet box is grounded by the bare ground wire of the cable, and the white wire, like all neutral wires, is grounded at the service panel.

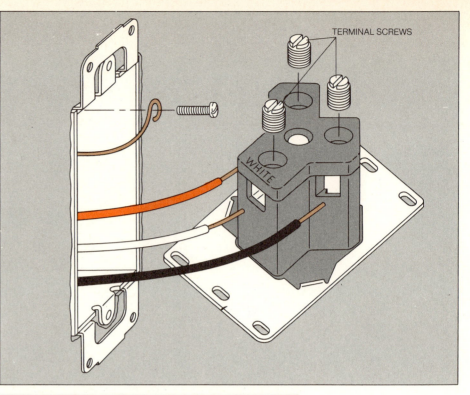

A 120/240-volt plug. Detach the plug cover and loosen the cord clamp. Push the cord under the clamp and into the plug. Attach the white wire to the terminal stamped "W" and the black and red wires to the other terminals. Tighten the cord clamp and remount the plug cover. In Canada, this circuit requires a fourth grounding wire and a fourth prong for the plug.

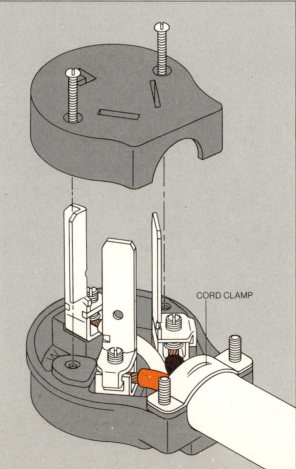

Surface-mounted receptacles. All 240-volt and 120/240-volt receptacles come in surface-mounted models, for use in basements and kitchens where in-the-wall wiring may not be practical. To remove the cover and expose the terminals, detach the mounting screw. Then connect the receptacle by the method that is used for a wall receptacle of the same voltage and amperage rating (*opposite top and below*).

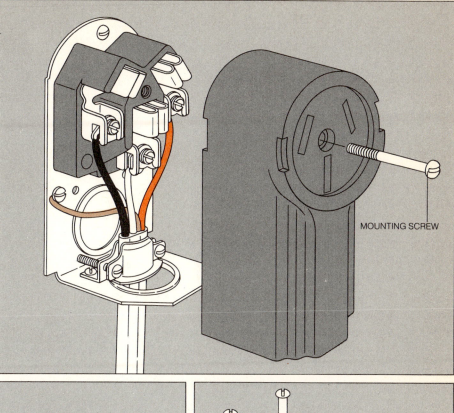

MOUNTING SCREW

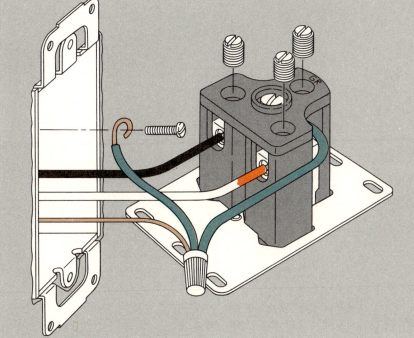

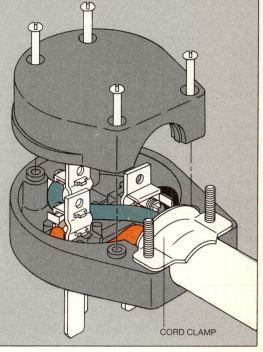

CORD CLAMP

A 240-volt, 30-ampere receptacle. This type needs no neutral wire, since 240 volts are provided by plug connection to both hot wires of a 240-volt cable. The cable may have only black and white wires—each to be connected to one brass terminal—plus a bare ground wire. Connect the bare wire to two green jumpers, one screwed to the back of the box, the other attached to the green receptacle terminal. Recode the white wire.

A 240-volt plug. Since this plug supplies only 240 volts, it needs no neutral wire. The black cord wire goes to one brass terminal, the white (or red) wire to the other brass terminal. If the cord has a white wire, recode it black. Attach the ground wire, which may be bare or green insulated wire, to the green terminal.

Cable for new wiring. A coil of plastic-sheathed cable awaits installation in a new circuit. This inexpensive cable is light and flexible enough to snake behind walls and ceilings. The cable can be cut and prepared for wiring with a wire stripper *(foreground)* and a utility knife, and it is easily attached to an outlet box *(right)*, which will house a new receptacle or switch.

No matter how large or versatile your wiring system, it is likely to be outmoded quickly. New appliances overcrowd existing receptacles, new room arrangements call for new lighting fixtures and switches. Yet straightforward procedures (more dependent on handyman carpentry than electrical know-how) enable you to add receptacles near plug-in appliances, put in three-way switches to turn lights on and off from two different locations, or install built-in fixtures—track lights, illuminated cornices or valances, or a whole luminous ceiling —to get the kind of lighting effects you see only in expensive homes.

To create these amenities you must run cables and install outlet boxes. This additional wiring is simple to conceal inside walls and ceilings so long as you work on interior partitions and ceilings. In most houses, ceilings and the walls between rooms are hollow so that you can snake cables inside them and sink boxes into them. The visible surfaces—plaster, plasterboard or paneling—are simply skins covering beams: vertical ones, called studs, in walls; horizontal ones, called joists, in ceilings and floors. The joists meet the studs at "plates," horizontal pieces topping the studs. This framework construction leaves empty space between skin surfaces for your wiring. Exterior walls, however, may be solid masonry, with no hollow spaces for snaking wires and sinking boxes. Even if they are made of wood and framed like interior partitions, the spaces between studs are likely to be insulated and are almost certain to be interrupted by fire stops, horizontal wood pieces inserted between studs to block the spread of flame. To extend wiring across an exterior wall, consider partially concealed wiring attached to the surface *(page 101)*.

So long as you work inside the hollows of floors, ceilings and interior walls, new cables can be run along or across studs, plates and joists from an existing outlet box to a new one. Usually, you must make holes in the skin surfaces—tiny test holes to locate studs and joists and larger access holes through which to snake new cable to gain access to the hollow interiors. Punching holes in your house is scary until you have done it a few times and discovered the holes will not cause permanent damage and are easy to patch.

The cable you run will normally be plastic sheathed *(left)*. It may not match older cables installed previously, but it can be used safely with any type of cable. If the old cable is sheathed with metal, you may not find a bare ground wire to tie into the new cable. Simply connect the bare wire of the new cable directly to the inside of the box, for the box itself provides a safe ground connection *(pages 18-19)*. At the new box, follow the procedures outlined in Chapter 2 for wiring plastic cable to the new lights, receptacles and switches that can modernize your home.

Planning a Circuit Extension Step by Step

No matter how simple or how ambitious a job you plan—installing a single receptacle or putting in an extensive circuit with several outlets—the number of steps and the order in which they should be done is the same in every type of home. The job goes much faster if you follow this logical sequence for putting in new boxes, running cable and connecting the cable to an existing circuit.

☐ Choose a location for the new box and break through the ceiling or wall to make an opening for it *(pages 64-75)*. Find the nearest joists or studs and, if the box is to be in a ceiling, determine the direction in which the joists run.

☐ Select an existing box as near the location of the new box as possible to provide power for the new circuit. The best starting point is a box that enables you to run cable to the new box on a direct route between the floor or ceiling joists, but alternative routes may be needed *(pages 76-83)*, depending on such factors as the direction of ceiling joists.

☐ Shut off the power at the service panel and remove the cover plate of the existing box. Check the wiring inside to see if you can connect the new cable wires to the existing wires or terminals. The insets on the opposite page show the wiring in typical home boxes that can be used to start new circuits. The wires represented as dash lines show how cable for the new circuit would be connected. You cannot bring power from a box that contains an end-of-the-run switch; and you should not bring power from the box of a light fixture controlled by a wall switch rather than a switch in the fixture itself, otherwise the added wiring will also be controlled by the wall switch.

☐ Count the connections in the box to see if there is room for more cable wires and clamps *(page 64)*. If the box is too small you must gang it to another box, install a deeper box or start the circuit from another place. Check the fuse or circuit breaker at the service panel to be sure that the existing circuit has enough capacity to run all the lights or appliances you may use in the new branch.

☐ Make a map of the room showing the existing box, the new box, the location of the studs and joists, and the route along which you plan to run cable. At the same time, make a list of the materials you will need. Start with a new box or boxes of the right size and shape, and the hardware for installing them.

☐ Estimate the amount of cable you will need to reach the new box. Provide an extra 8 inches of cable for connections at each box and an extra 20 per cent to allow for the fact that the cable will rarely run perfectly straight. Choose cable that is the correct gauge for the circuit *(page 24)* and contains the right number of conductors. The sample circuits that are shown on page 13 illustrate situations in which you will require two-conductor and three-conductor cables for the different parts of a run. Finally, list the type and number of clamps that will be needed to attach the cable to both old and new boxes *(pages 84-85)*.

☐ Run the new cable from the existing box to the opening for the new one. This is usually the most strenuous part of the job. It may involve cutting access holes in walls or ceilings, drilling a path for the cable through beams along the cable's route and snaking the cable through the walls or ceilings *(pages 76-83)*.

☐ Clamp the cable to both the existing box and the new box. Normally, the existing box is permanently installed in a wall or ceiling; therefore, the cable must be attached with an internal clamp. The new box, which has not yet been installed, will accept an internal or an external clamp.

☐ Install the new box. The procedure you use will depend on the composition of the wall or ceiling and on whether the box can be mounted directly to a stud or joist *(pages 64-75)*.

☐ Connect the wires of the cable to the correct wires or terminals in the existing box *(opposite)*, and then also to the switches, receptacles or light fixtures in the new boxes.

☐ Test the new circuit branch. First, clip a continuity tester *(page 10)* to the black wire in the box at the end of the run, and touch the probe to the white wire, the bare wire and to the box. The tester bulb should not glow in any of these tests. If it does, there is a short circuit in the new branch: check at each new box for improper connections, frayed insulation or a bare wire touching another wire, and make any corrections necessary. Then go on to a second set of tests. Have a helper turn on current to the circuit at the service panel and use a voltage tester to check for power in each device on the new circuit *(page 22)*. After these tests, turn off the power immediately.

☐ Patch up all the holes you have made in the walls or ceilings *(pages 104-105)*. Then turn on the power again. Your new branch is ready for use.

Wiring an Old Outlet Box with New Cable

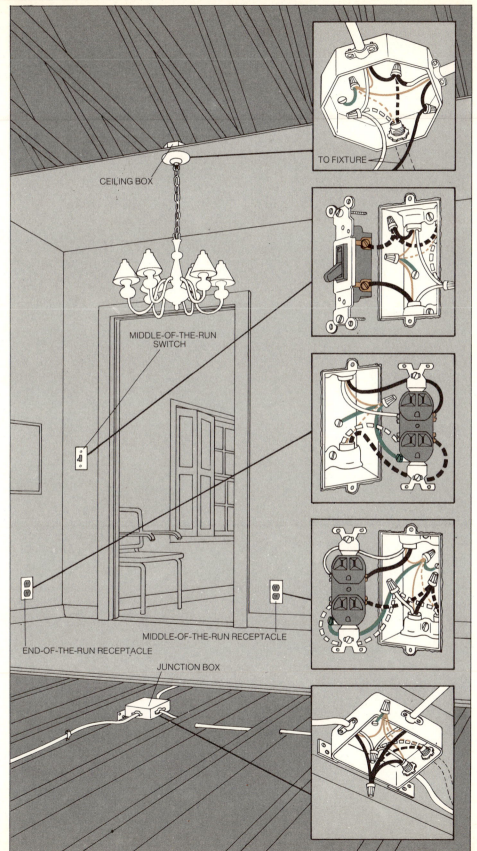

CEILING BOX

MIDDLE-OF-THE-RUN
SWITCH

TO FIXTURE

MIDDLE-OF-THE-RUN RECEPTACLE

END-OF-THE-RUN RECEPTACLE

JUNCTION BOX

Middle-of-the-run ceiling box. Make sure that the ceiling box is middle-of-the-run. If it is, the black and white fixture wires will be connected to two separate cables. Otherwise, current will flow in the new branch only when the switch controlling the ceiling box is on. If the box is middle-of-the-run, use a voltage tester to locate the feed cable coming from the service panel (*page 22*). Attach the wires of the new cable (*dash lines*) to the incoming black, white and bare wires.

Middle-of-the-run switch. Determine that the switch is a middle-of-the-run power source—it will be connected to two separate cables. If it is, use a voltage tester to identify the black wire that carries current from the service panel (*page 22*). Disconnect it and join it to the black wire from the new cable and to a black jumper (*dash lines*). Attach the jumper to the switch terminal. Join the new white and bare wires (*dash lines*) to the white and bare wires in the box.

End-of-the-run receptacle. This is the easiest location from which to extend wiring—as long as the receptacle is not controlled by a switch. Attach the black and white wires of the new cable (*dash lines*) to the unused terminals on the receptacle (*page 54*). Join the bare wire to the existing bare wire and the jumper with a wire cap.

Middle-of-the-run receptacle. To accommodate the additional wires and clamps, you probably will have to gang the existing box (*page 65*). Remove a black wire from a receptacle terminal and join it with both the black wire of the new cable and a black jumper wire (*dash lines*). Attach the jumper to the terminal. Use the same procedure for a white receptacle wire and the white cable wire. Attach the new bare wire (*dash lines*) to the other bare wires and the green jumper wire.

Junction box. Since the wires of more than one circuit may pass through a junction box, first locate the wires of the circuit you plan to tap for a power source. Have a helper shut off the circuits at the service panel one by one until you locate, with a voltage tester, the wires controlled by the fuse or breaker in the circuit you wish to tap (*page 22*). Then shut off the circuit and attach the black, white and bare wires in that circuit to wires of the same colors in the new cable (*dash lines*).

Setting Outlet Boxes into Walls and Ceilings

The central element in any permanent electrical installation is an outlet box—a metal or plastic enclosure embedded in a wall or ceiling. A receptacle, switch or light fixture is mounted inside a box, cables enter and leave through holes in the back or sides, and all wiring connections are entirely contained within the box.

Boxes vary in shape according to their position and function: rectangular wall boxes are used with receptacles, switches and wall lights; octagonal or circular ceiling boxes are used with ceiling lights; and square junction boxes, the largest of all, house connections between two or more cables that branch out to different parts of a circuit.

The existing boxes in your home, installed when the house was built, are probably attached directly to studs (the vertical posts that support a wall) or to joists (the beam to which a ceiling is fastened). Plastic boxes, widely used in houses built since about 1970, are always installed in this way. When you install new boxes, however, you probably will not have direct access to studs or joists—they are covered over by walls and ceilings. Therefore you must use a metal box secured in other ways. The method you choose depends upon the type of construction in your house.

In a house built after 1950, walls and ceilings are most commonly constructed with plasterboard—4-by-8-foot sheets of plaster that are sandwiched between two outer layers of paper and nailed directly to the studs or joists. Other wall and ceiling materials include wood panels, plywood sheets or sheets of compressed wood fibers called hardboard. Like plasterboard, all these wooden walls are fastened directly to the studs or joists.

A common surface in homes constructed before 1950 and in many of the newer apartments is plaster. It is applied either to wood lath (strips of wood about 1½ inches wide, nailed across the studs and joists about ½ inch apart) or to metal lath (a strong mesh that is made of interwoven metal strips). Instructions for installing wall boxes in all these surfaces begin on page 66; instructions for installing ceiling boxes appear on pages 70-75.

Before you install a new box, locate an existing box from which you can run new cable. Make sure that the existing box can safely accommodate additional wires *(chart, below)*. If the box has reached its capacity and you cannot find a second box from which to run new cable, you can enlarge a wall box *(opposite, bottom)* or replace a box with a deeper one.

How Big a Box to Use

Use the chart at right to determine how many wires you can safely put in an outlet box. Measure the depth of the box, determine the size of the wires already installed or to be added *(page 24)*; check the chart for the maximum number of connections permitted.

The capacity of a box depends not only on the number of wires inside it but also on the number of devices that take up space. Count as one connection each wire, except ground wires, that enters the box; do not count jumper wires. Add one connection if the box contains any ground wires inside it, another if the box contains any internal cable clamps or light-fixture studs, and still another if it contains any switches or receptacles. Then check the total against the number of connections listed in the chart.

A standard 2½-inch-deep wall box might contain a cable with two No. 14 conductor wires and a ground wire inside, plus an internal clamp and receptacle. The box has five connections; its maximum capacity, according to the chart, is six. Though many electricians would use this box as a starting point

for a new cable, the National Electrical Code ruling is clear: Before new wiring is installed, the box should be replaced with one that is deeper or ganged with a second box of the same depth *(opposite, bottom)*.

In an older home, the same 2½-inch box might contain a metal-armored cable attached by an external clamp (not counted as a connection). Here the connection count would be only four—two conductors, plus a ground wire and a receptacle. A second cable containing two No. 14 conductor wires and a ground wire can be safely added by using a two-part external clamp.

Matching Connections to Box Size

	Box depth	Maximum number of connections No. 14 wire	Maximum number of connections No. 12 wire	Maximum number of connections No. 10 wire
Wall boxes	2½''	6	5	5
	2¾''	7	6	5
	3½''	9	8	7
Ceiling boxes	1¼''	6	5	5
	1½''	7	6	6
	2⅛''	10	9	8
Junction boxes	1¼''	9	8	7
	1½''	10	9	8

The Three Types of Wall Boxes

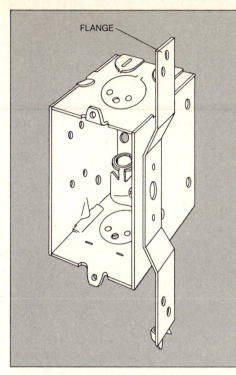

FLANGE

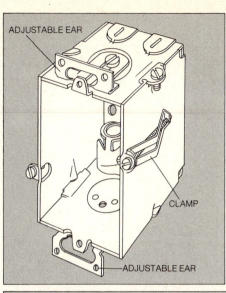

ADJUSTABLE EAR

CLAMP

ADJUSTABLE EAR

A box with side clamps and ears. In plywood paneling or hardboard walls, this box is secured in an opening by clamps that expand behind the wall when the screws at the sides of the box are tightened. The adjustable ears projecting from the top and bottom of the box prevent it from falling backwards into the wall space (*page 66*).

A box with ears and optional brackets. In a plaster wall with a wood-lath backing, which holds screws well, this wall box can be secured by its adjustable ears alone. For plasterboard and for plaster over metal lath, neither of which holds screws well, use the brackets shown here and follow the mounting method described on page 67. The brackets are purchased separately.

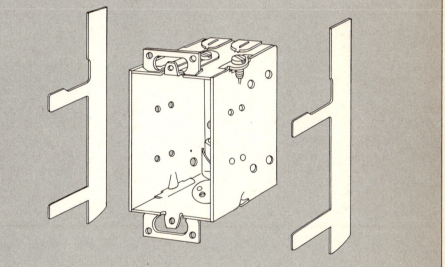

A box with a flange. Nailed to a stud by its flange, this type is the easiest wall box to install and is probably the one already inside the walls of your house. Although you cannot add one where the walls are finished, it is a first choice for a basement or attic with exposed studs. If you plan a finishing surface over the stud, allow for its thickness; position the flange to bring the front of the box flush with the surface to be added.

Enlarging a Wall Box

Ganging boxes. Two or more wall boxes of the same depth can be ganged to multiply their size so they can hold extra wires and more than one switch or receptacle. Loosen the screw in the flange at the bottom of one box and pull off the left side. Remove the right side of the other box by loosening the screw at its top. Then hold the open sides together, positioning the notch on one box between the screwhead and flange of the other, and tighten the two screws.

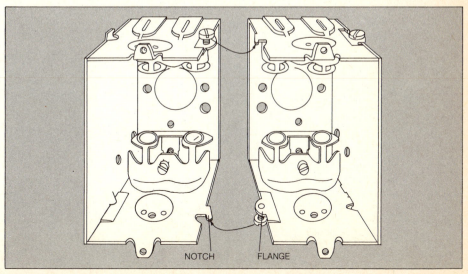

NOTCH   FLANGE

Boxes in Wood Walls

1 Locating the box. Mark the wall approximately where you want the box and drill a ⅛-inch hole through the wall at this point. If the bit hits a stud, fill the hole with spackling compound and drill a second hole about 6 inches to the left or right. Bend an 8-inch-long piece of coat-hanger wire to a 90° angle at the middle, insert the wire into the wall and turn it through a complete circle. If it does not turn freely within the wall space, indicating the absence of pipes or framing, cover the second hole and repeat the procedure a few inches away until you locate a clear wall space.

2 Using a template. Place a box with side clamps face down on a sheet of thick paper. Trace the outline of the box on the paper with a pencil; do not include the detachable ears at the top and bottom of the box. Cut out the paper within the outline as a template and center the template on the hole you have drilled in the wall, with the side of the paper on which you traced the outline against the wall. The template now indicates the exact positions of each of the side clamps. Mark the outline of the template on the wall with a sharp-pointed tool such as an awl.

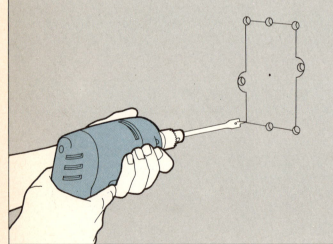

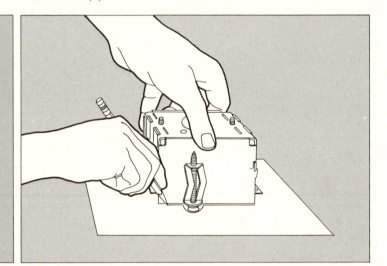

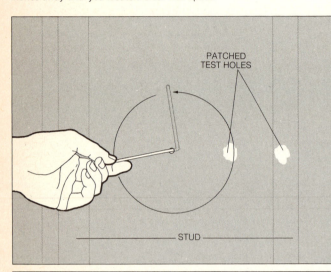

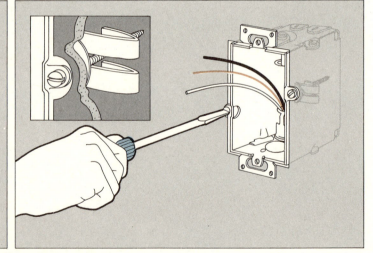

3 Cutting a hole. Bore a starter hole for a keyhole saw at each of the four corners on the outline, positioning the point of the bit just outside the outline. Then bore two holes for the side clamps on the box, and two more at the top and bottom of the outline to provide clearance for the long screws that will fasten a switch or receptacle to the box. Cut around the outline with a keyhole saw. Use short strokes so you do not jab the blade of the saw into the other side of the wall.

4 Installing the box. Push the box into the wall until the ears hold the front edge flush to the surface of the wall; if necessary, loosen the ear screws and adjust the positions of the ears. Remove the box and attach cable through a knock-out hole (pages 84-85). Then, holding the box in place, tighten the screws on the sides. Behind the wall, the side clamps will bulge outward, drawing the box into the wall and forcing the ears firmly against the outside surface.

Boxes in Plasterboard

1 **Cutting an outline.** Follow Steps 1 and 2 opposite, using a standard wall box as a guide for the template. Cut along, and just outside, the template outline with a utility knife. Place the point of the blade at one of the top corners of the outline. Grip your right wrist in your left hand, bracing your left elbow against the wall, then cut downward. Guide the knife with your right hand while your left hand pushes and steadies the right. Go around the outline twice to cut completely through the paper on the room side of the plasterboard; you need not cut through the board.

2 **Breaking through the plasterboard.** Cut a piece of scrap wood about 3 inches long, 2 inches wide and 1 inch thick. Place one end at the center of the outline you have cut on the wall and strike the other end with a hammer. Try to hit hard enough to break through the paper on the back of the plasterboard. If all of the outlined section does not break away, hammer the piece of wood along the edges of the outline until you have made a rectangular opening of the right size. Then cut away the uneven fragments of plasterboard and paper with the utility knife.

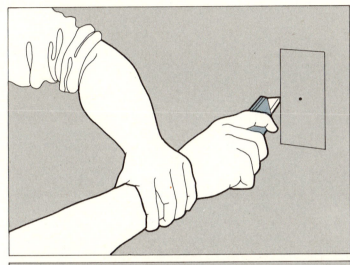

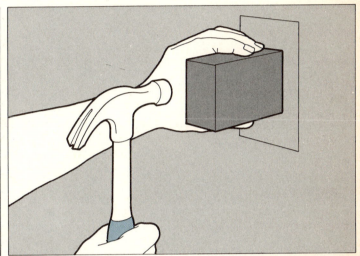

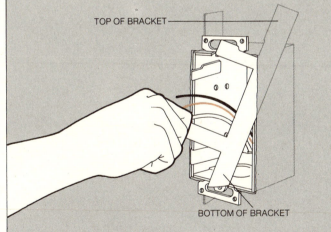

TOP OF BRACKET

BOTTOM OF BRACKET

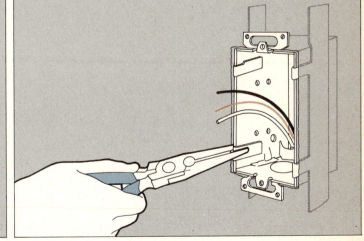

3 **Installing the box.** Attach cable to the box (pages 84-85). Adjust the box ears to bring the front of the box flush with the wall, then push the box into the wall and hold it in place with one hand. With your other hand, insert a bracket between one side of the box and the edge of the opening—slide the top of the bracket in first, then the bottom. Caution: Hold the bracket tightly; if it drops behind the wall, it is irretrievable. Pull the bracket toward you by its arms as far as you can, then bend the arms into the box. Install a bracket on the other side of the box in the same way.

4 **Tightening the brackets.** Grip the bend in one of the bracket arms with needle-nose pliers and squeeze it tightly against the inside of the box. Repeat this procedure with the other three bracket arms. The arms should be as tight as possible, not only to secure the box firmly, but also to keep the arms out of the way of the wires.

Boxes in Walls of Plaster on Wood Lath

1 Exposing a lath. Find a clear wall space for a box by drilling a hole and following the procedure on page 66, Step 1. Using a cold chisel and ball-peen hammer, remove plaster around the drilled test hole, working upward and downward until you have exposed the entire width of a single lath—normally about 1½ inches is enough.

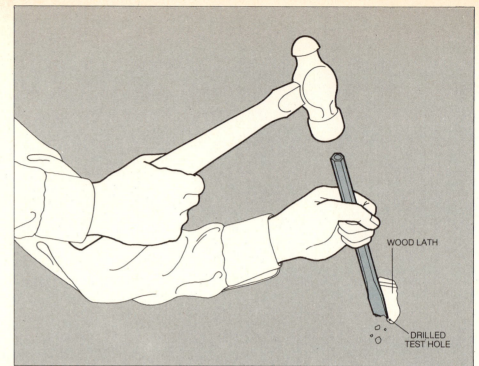

WOOD LATH

DRILLED TEST HOLE

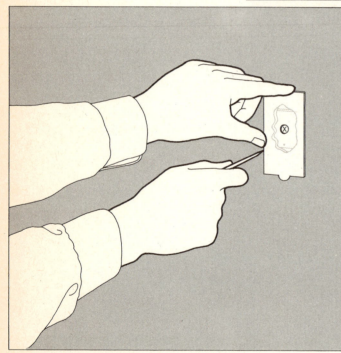

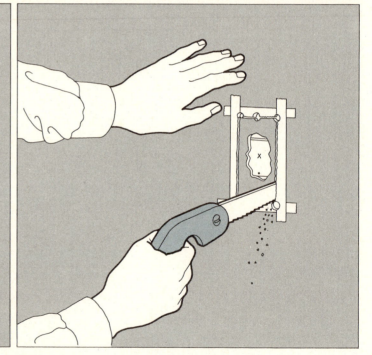

2 Drawing the outline. Mark a penciled dot exactly halfway between the top and bottom of the lath. Cut a template for the box (*page 66, Step 2*), using a standard wall box rather than a box with side clamps. Cut a small hole at the exact center of the template and hold the template against the wall with the hole directly over the mark on the lath. Trace the template outline.

3 Cutting through the wall. Apply strips of masking tape around the outline to keep plaster from breaking. Score the outline several times with a utility knife. Bore ⅜-inch holes at the outline corners and in the curves at the top and bottom; then cut along the scored lines with a keyhole saw. Work smoothly and evenly; irregular strokes of the saw can jar the laths and crack plaster.

4 **Installing the box.** Remove the masking tape, insert the box in the wall opening and trace the outlines of the ears. Then remove the box and score these outlines with a utility knife. Using a cold chisel and ball-peen hammer, cut away the plaster within the outlines to expose the laths beneath. Chisel in from the outlines toward the wall opening to minimize damage to the wall.

Hold the box in the wall opening and mark the positions of the screw holes in the ears on the exposed lath; drill pilot holes at these points for the mounting screws. Position the movable ears to bring the front of the box flush to the wall surface. Then attach cable to the box (*pages 84-85*) and screw the box to the lath.

WOOD LATH

Boxes in Walls of Plaster on Metal Lath

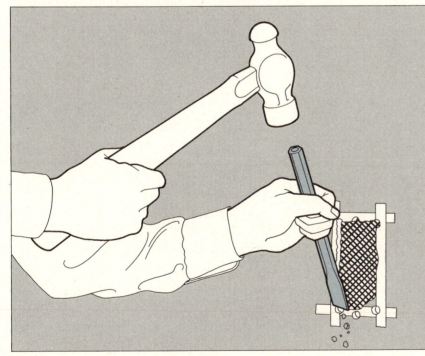

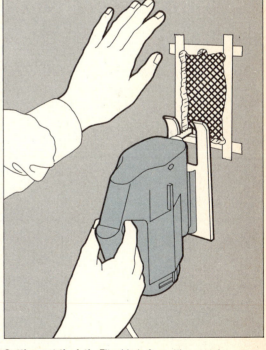

1 **Chiseling out the plaster.** Locate a clear space for the box and trace the outline of the box on the wall (*page 66, Steps 1 and 2*), using a template for a standard wall box rather than a box with side clamps. Tape and score the outline (*opposite page, Step 3*). Drill holes at the four corners of the outline and in the curves at the top and bottom, using a ⅜-inch metal bit. With a cold chisel and a ball-peen hammer, remove all of the plaster within the outline so that the metal lath beneath is completely exposed.

2 **Cutting out the lath.** Fit a blade for cutting metal to a saber saw, insert the end of the blade in one of the holes that you drilled at the edges of the outline and cut out the exposed metal lath. Caution: This part of the job must be performed very slowly and with special care to prevent the vibration of the saw from damaging the plaster around the opening. Install a standard box, using brackets and following the procedure described for installing wall boxes in plasterboard (*page 67, Steps 3 and 4*).

Installing Boxes in Ceilings

A ceiling box may be mounted in any of three ways. It can be attached directly to a joist, it can be suspended between two joists by a device called a bar hanger or it can be hung from the bottoms of joists by a special type of bar hanger called an offset hanger. Your choice among these boxes will depend partly on the material of which the ceiling is made and partly on the access you have to the space above the ceiling.

The ceiling material you are most likely to encounter will be either plasterboard or, in older houses, plaster over wood lath. Both are easy to work with when you have access to the ceiling from an attic above it. If the attic has floorboards, you can remove part of a board to get at the joists. If the joists are exposed, the installation job is even simpler but calls for one special precaution. Though the joists themselves will support your weight, the ceiling below them will collapse if you step on it. You should create a patch of sturdy flooring by laying three or four planks across the joists at your work area.

If the ceiling lies directly under the roof or under another room, you must install the box from below. When you work from below, the ceiling material is crucial; the procedure for a plasterboard ceiling is quite different from the one for plaster (pages 73-75).

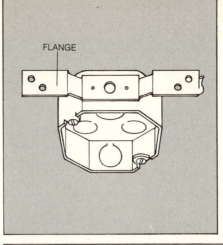

FLANGE

Three Types of Ceiling Boxes

A box with a flange. Screws or nails inserted in the permanently attached flange fasten this box directly to a joist. The installation is simple but the box suffers from one limitation: it can be installed only at the side of a joist, not in the comparatively large spaces between joists.

A box with a bar hanger. This box is screwed to the bar hanger with a threaded stud, the sliding arms of the hanger are extended to meet the joists on either side, and the entire assembly is fastened to the sides of the joists by the flanges at the ends of the arms. Because the construction of this box permits it to be positioned at almost any point between two joists, the bar hanger type has become the most popular.

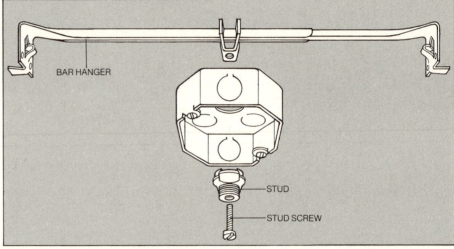

BAR HANGER
STUD
STUD SCREW

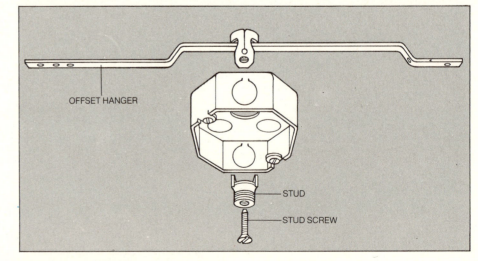

OFFSET HANGER
STUD
STUD SCREW

A box with an offset hanger. This device is specially designed to be installed only in a plaster-and-wood-lath ceiling that cannot be reached from above. In this situation, it is impossible to attach a flange to a joist, or a bar hanger between two joists. The offset hanger fits into a groove cut into the plaster of the ceiling and is fastened to the exposed bottoms of two joists (page 75).

Ceiling Boxes Positioned from an Attic

1 **Locating the box.** Choose a place for the new ceiling box and mark the spot on the surface of the ceiling with a pencil or awl. Drill a ⅛-inch hole through the mark. The bit should penetrate the ceiling almost immediately. If it meets prolonged resistance, you are drilling into a joist; cover the hole with spackling compound and drill a second hole at least 2 inches away. When you have found a clear space, bore a ¾-inch hole, then fit an 18-inch-long extension and a ⅛-inch bit to the drill and bore a hole through the attic floorboard directly above. Go to the attic for the next step. (If your attic does not have floorboards, proceed to Step 3.)

2 **Removing a floorboard.** In the attic, find the board with the drilled hole in it. The joists beneath this board can be located by the rows of nails that fasten the board to the centers of the joists. The edges of the joists lie about 1 inch from the rows of nails. Mark the joist edges that are nearest to the hole, using a steel square and an awl to score lines across the floorboard. Bore a ⅜-inch starter hole at one end of each of the lines you have marked; then cut through the board along the lines with a saber saw or keyhole saw. Retrieve the loose section of board you have sawed free, so that you can replace it in the opening when the job is completed.

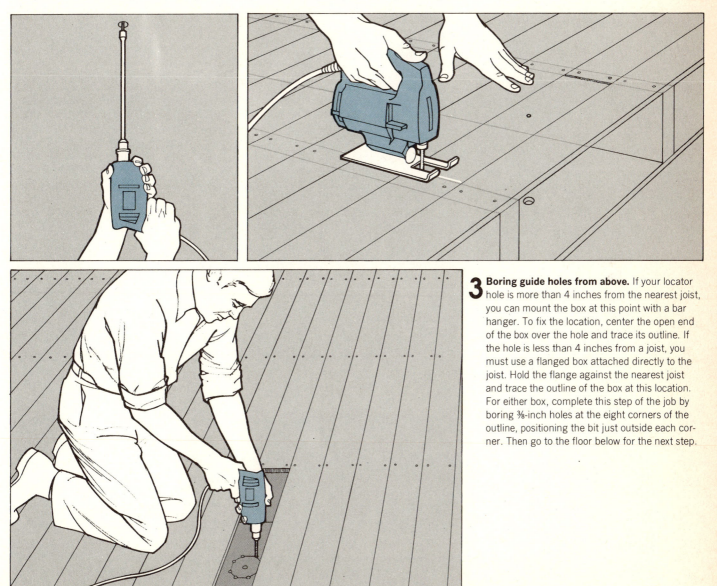

3 **Boring guide holes from above.** If your locator hole is more than 4 inches from the nearest joist, you can mount the box at this point with a bar hanger. To fix the location, center the open end of the box over the hole and trace its outline. If the hole is less than 4 inches from a joist, you must use a flanged box attached directly to the joist. Hold the flange against the nearest joist and trace the outline of the box at this location. For either box, complete this step of the job by boring ⅜-inch holes at the eight corners of the outline, positioning the bit just outside each corner. Then go to the floor below for the next step.

4 **Cutting the box hole from below.** Wearing a pair of protective goggles and using a keyhole saw, cut out the outline marked by the drilled holes. A plaster-and-lath ceiling calls for special precautions: score the plaster with straight lines between the holes; apply masking tape around the outline *(page 68, Step 3)*; and brace a piece of wood against the ceiling just outside the lines of the outline as you make your cuts. A plaster-board ceiling is easier to work with; simply brace the ceiling with your hand as you saw. In either material, begin the cuts by ramming the point of the keyhole saw into one of the holes and work from hole to hole around the entire outline.

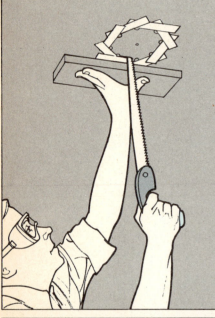

5 **Installing the box.** Install the cable into the box. If you are working with a bar hanger, fasten the box to the hanger with a threaded metal stud and cut the small tabs from the flanges at the ends of the hanger with metal snips. Holding the box in the ceiling hole with its front flush to the lower surface of the ceiling, extend the arms of the hanger to the joists on either side. Mark the positions of the hanger screw holes on the joists, bore pilot holes for the screws and screw the hanger to the joists *(left)*.

To install a flanged box, hold the box with its front flush to the lower surface of the ceiling and the flange against the side of the joist. Mark the positions of the flange screw holes on the joist, drill pilot holes for the flange screws and screw the flange to the joist.

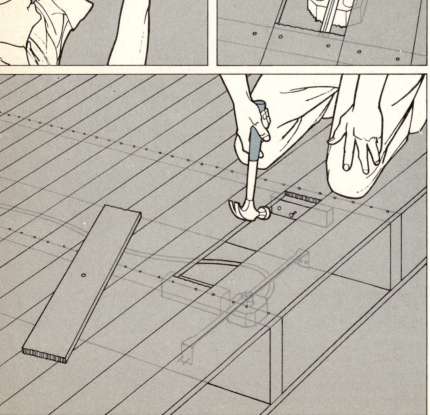

6 **Replacing the board.** When the cable and the new box have been installed, cut two wooden cleats, each 2 inches wide, 3 inches high and slightly longer than the width of the board you have removed. Nail the two cleats to the exposed sides of the joists. Then insert the loose board in the opening and nail it to the tops of the cleats.

Ceiling Boxes Positioned from Below: in Plasterboard

1 Measuring the plasterboard. Drill a ⅜-inch hole in the ceiling at the point you have chosen for the box and use a piece of bent wire to be sure that the space above is clear *(page 66, Step 1)*. With a keyhole saw, cut a roughly square opening about 6 inches wide around the drilled hole. Insert a steel tape into the opening and measure the distance from the edge of the opening to the side of an adjoining joist. Add 1 inch to this distance and mark off the total distance on the outside of the ceiling. The mark will lie just beyond the center of the joist. Repeat this measurement from a second point on the edge of the opening and mark it. Score a line through the two marks, using a carpenter's square as a guide. Repeat the procedure at the opposite edge of the opening, measuring the distance to the joist on the other side. Connect the lines to form a 16-inch square with the opening at its center.

2 Cutting into the board. Set the steel square at one of the inside corners of the area you have scored and cut into the plasterboard along both sides of the square with a utility knife. Be careful not to cut beyond the corners of the scored area. Cut into the other sides of the scored area in the same way. Then make four cuts from the corners of the scored area to the opening at its center.

3 Breaking the board away. Wearing protective goggles, break up the plasterboard within the square with a hammer. Use a utility knife to cut away the inner layer of paper holding the broken fragments of plasterboard to the ceiling.

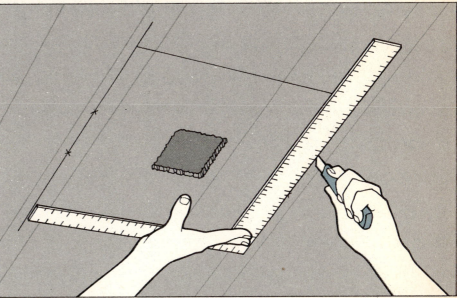

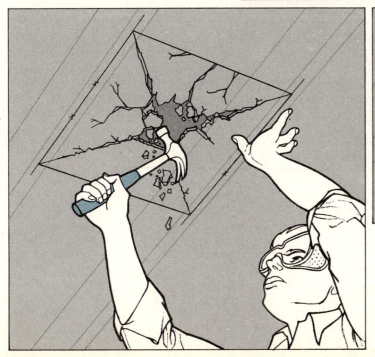

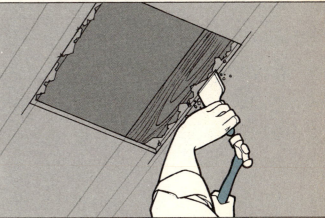

4 Trimming the edges of the square. Use a chisel—the wide-bladed type called a brick chisel or brickset works best—and a ball-peen hammer to remove any plasterboard that clings to the joists. Set the chisel blade along the edge of the board and hammer through the plaster and paper to the joist itself. Pull the nails from the joists with a hammer, bracing the hammer against the joist rather than the delicate plasterboard.

5 **Cutting new plasterboard.** Place a panel of plasterboard on a flat surface. Measure the exact dimensions of the square you have cut in the ceiling, subtract ⅛ inch from each side of the square and draw the slightly smaller square on the plasterboard panel. Using the steel square as a guide, cut through the upper layer of paper along one side of the square with a utility knife, then extend the cut to the edges of the panel. Place the panel on a workbench with the cut directly above the edge of the bench. Hold the panel firmly on the bench with one hand and hit the area beyond the cut sharply with the palm of the other; the panel should snap along the cut. Cut through the bottom layer of paper and repeat the procedure on the other sides of the square.

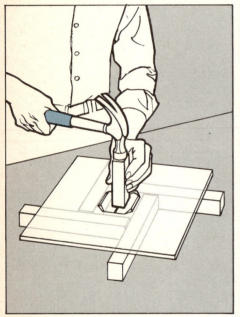

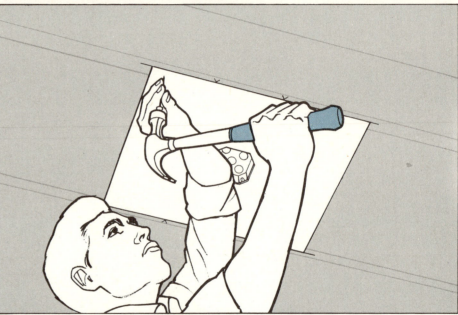

6 **Making an opening for the box.** Trace the outline of the ceiling box at the center of the plasterboard square. Cut along the box outline with the utility knife and make an additional cut through the middle of the outline. Prop the plasterboard square on four pieces of wood and hammer a block of wood through the center of the outline; cut away the plasterboard fragments. Install the bar hanger and the ceiling box (*page 72, Step 5*), then check that the box lies directly above the hole you have just made. If necessary, adjust the position of the box on the hanger or move the hanger to another spot on the joists. Attach cable to the box (*pages 84-85*).

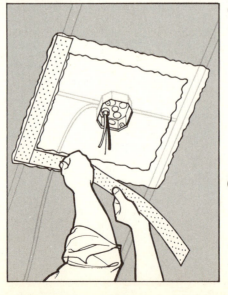

7 **Installing the plasterboard square.** Brace the plasterboard square, dark side up, against the exposed joists with one hand and forearm. Hold a plasterboard nail at a corner of the square with that hand, and hammer the nail through the plasterboard and into the joist with the other hand. Drive the nail flush to the surface of the plasterboard, but be careful not to break the paper surface. Repeat this procedure at the other three corners, then secure the section to the joists with a series of nails spaced about 3 inches apart.

8 **Completing the job.** Fill the cracks at the sides of the square with plasterboard joint cement and cover the cracks with strips of perforated joint tape (the strips should not overlap). Spread a thin coat of the cement over the tape, smooth the surface and let the compound dry. Then apply heavier coats of joint cement until the tape is completely covered. You may need three or four coats. Sand the final coat smooth and paint it.

Ceiling Boxes Positioned from Below: In Plaster

1 **Locating the joists.** Wearing protective goggles and using a ½-inch cold chisel and ball-peen hammer, remove enough plaster at about the point you plan to install the box to expose the complete width of one lath. If you find nails in the lath, a joist lies directly above. Chisel along the lath in either direction until you reach a second joist. If you do not find nails at the starting point, chisel along the lath until you do, then go back to the starting point and chisel along the lath in the other direction to the next joist. Trace the outline of the box at the center of the channel and cut an opening for the box *(page 72, Step 4)*.

2 **Installing the box and hanger.** Chisel out enough plaster under the joists to expose their bottom edges. Cut the two exposed sections of lath at the outside edge of each joist with a keyhole saw and remove the nails from the joists. Attach the box to an offset hanger, hold the box in the opening with the arms of the hanger against the joists and mark the positions of the hanger screw holes on both joists. Drill pilot holes into the joists for the hanger mounting screws. Then attach the cable to the box *(pages 84-85)* and fasten the hanger to the joists with 1½-inch wood screws.

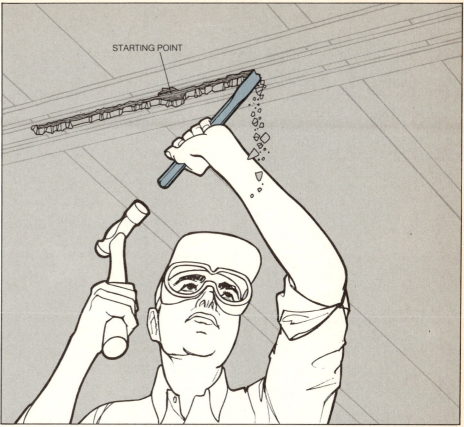

STARTING POINT

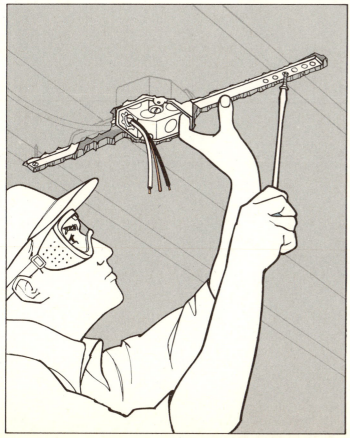

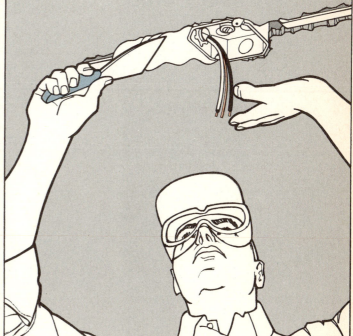

3 **Covering the channels.** Fill the two channels with patching plaster to a level just below the ceiling surface. After the plaster sets, fill the channels flush with spackling compound or plasterboard joint cement. When dry, sand smooth and paint.

Running Cable from Old Outlet Boxes to New

Having made an opening for a new outlet box—but before installing the box—you must run cable to the opening from an existing box. First find a nearby box to supply power for the new cable, enlarging it if necessary *(pages 64-65)*. Then select a route for the cable that calls for the least labor and causes the least damage to your walls and ceilings.

The best route on both counts runs to the maximum extent possible through an unfinished basement or attic, where wall studs and ceiling or floor joists are either exposed or easily uncovered. You can drill holes through the studs or joists *(page 78)*, then thread cable through the holes; electrical codes require only that the holes be drilled at least 1½ inches from the edge of a stud or joist, and that

you staple the cable to a support *(opposite, bottom)* within 1 foot of a box. Running cable along the side of a stud or joist is even easier: the only code requirement is that you staple the cable at regular intervals *(opposite, bottom)*.

Where a finished room lies above a basement or below an attic, these methods suffice for floor or ceiling outlets. A wall outlet needs another step: boring a hole through the plates, or beams, at the top and bottom of the wall. Through this access hole you can feed the cable inside the wall to the outlet hole.

Running cable in a finished room that has no access to a basement or attic calls for a radically different procedure. To begin with, you must make holes in a wall or ceiling to get at hidden studs, joists

and plates *(pages 79-83)*. Through these access holes you can bore holes through studs and plates, and then, with a fish tape, pull the cable behind a wall from stud to stud, or behind a ceiling between two joists. Other access holes enable you to run cable over a doorway, through a wall, or along a ceiling from an adjoining room *(pages 82-83)*.

Often, you will have to combine several methods of running cable. In the example on these pages, cable for a new wall outlet is connected to an existing wall outlet by running it up through the walls at either end and along basement joists in between. On the following pages are techniques for running cable across joists, through an attic and inside a wall inaccessible from basement or attic.

A Cable Route along Basement Joists

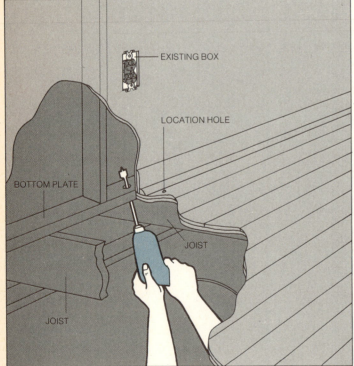

EXISTING BOX
LOCATION HOLE
BOTTOM PLATE
JOIST
JOIST

1 Getting from wall to basement. Bore a ¹⁄₁₆-inch location hole through the floor directly below the existing box you plan to tap. Poke a thin wire through so you can find the hole in the basement. Then, in line with the location hole, drill up through the plate with a ¾-inch spade bit.

Directly across the room from the old box, cut an opening for the new one and, using the same method, drill a hole through the plate below it.

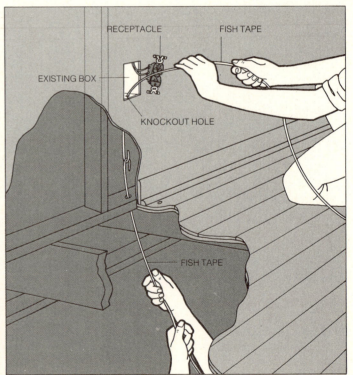

RECEPTACLE
FISH TAPE
EXISTING BOX
KNOCKOUT HOLE
FISH TAPE

2 Fishing tape through the wall. Detach the receptacle from the existing box and pull it out of the way—you need not disconnect the wires. With a hammer and nail set, remove a knockout from a hole in the bottom of the box *(pages 84-85)*. Push the end of a fish tape through this hole and down behind the wall, and have a helper push a second tape up through the hole in the bottom plate. Maneuver both of the tapes behind the wall until their ends hook tightly together.

3 **Attaching cable to a tape.** From the basement, pull down the end of the upper tape through the hole in the bottom plate; unhook the two tapes. Strip 3 inches of sheathing from the end of a cable and strip the insulation from the exposed wires. Run the ends of the bare wires through the hook of the upper tape and loop the wires back over themselves. Tape the hook of the fish tape and the looped wires firmly so the cable will not snag or pull loose from the fish tape.

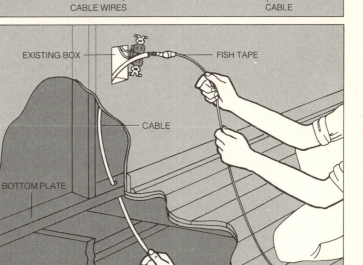

ELECTRICIAN'S TAPE

FISH TAPE

CABLE WIRES   CABLE

4 **Running the cable to the existing box.** From the room, pull the fish tape back through the knock-out hole in the existing box until the end of the cable emerges; to make this step of the job easier, have your helper feed the cable up through the bottom plate from the basement. Detach the cable from the fish tape. Strip 8 inches of sheathing from the end of the cable and fasten the sheathed end of the cable to the box with an internal clamp (*pages 84-85*).

EXISTING BOX   FISH TAPE

CABLE

BOTTOM PLATE

5 **Completing the job.** Run the cable across the basement ceiling, fastening it to the nearest joist with cable staples at 4-foot intervals. (Be careful not to damage the cable sheathing as you nail the staples.) When you reach the hole you have drilled in the opposite bottom plate, fish the cable up to the new box opening by the method shown in Step 2. Attach the cable to the new box and mount the box in the wall (*pages 64-69*).

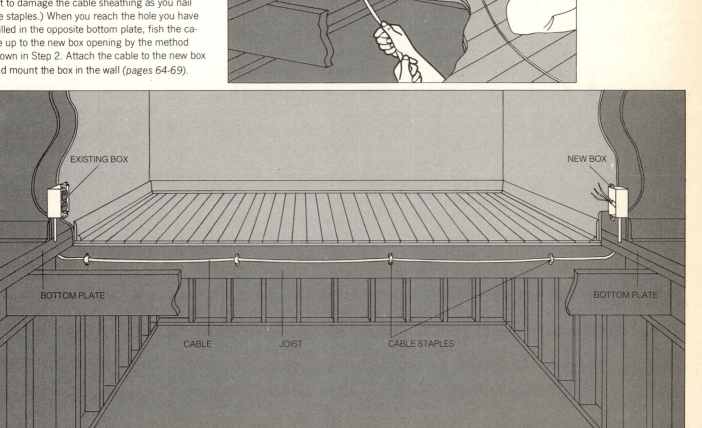

EXISTING BOX   NEW BOX

BOTTOM PLATE   BOTTOM PLATE

CABLE   JOIST   CABLE STAPLES

Alternate Routes in Basements and Attics

Running cable across basement joists. If your route for new cable runs at right angles to the basement joists, run the cable through ¾-inch holes bored through the middle of each joist. The holes should run in a straight line. Attach the cable to the existing box *(pages 76-77, Steps 1-4)* and snake the cable through the joists to a point directly below the new box, then fish the cable up to the opening and attach it to the box.

Fishing cable to an attic. Cable can be run along or across exposed attic joists by the same procedures that are used in a basement—except for a slight variation in fishing the cable.

Remove a knockout from a hole in the top of the existing box. Drill a hole through the top plates of the wall at a point directly above the box and have a helper drop a light chain down through the plate, far enough to reach below the box. Push a fish tape through the knockout hole, catch the chain and pull it into the box. Hook the fish tape to the end of the chain and tape the hook and chain together. Pull the chain up until the fish tape reaches the attic, then detach the chain, attach the cable to the tape hook as shown in Step 3 on page 77 and then proceed with Steps 4 and 5 on that page.

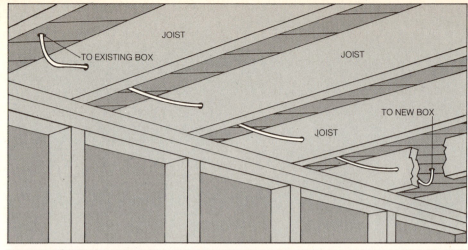

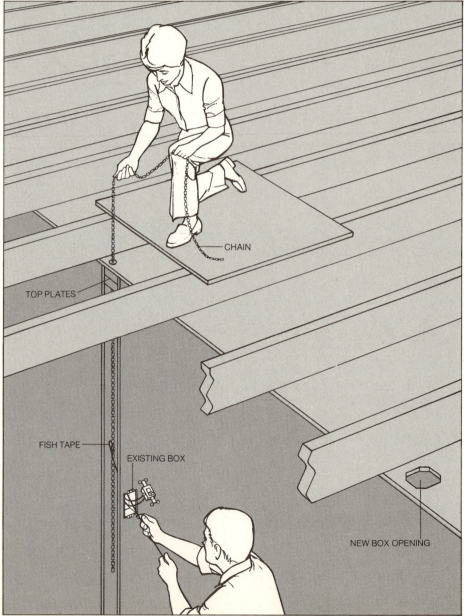

A Route behind a Wall

1 Exposing the studs. Make an opening for the new box. Drill a 1 1/16-inch test hole through the wall near the existing box and insert a stiff wire through the hole toward the new box until you touch a stud. Measure the length of the wire from the hole to the stud and mark the wall at that distance. Then drill additional test holes, if necessary, in order to locate the stud precisely.

If the wall is plasterboard, use a utility knife to cut a rectangle 3 inches high and wide enough to extend 2 inches beyond the edges of the the stud. Knock out the plasterboard rectangle to expose the stud (*page 73*). Locate the next stud and cut a hole over it by the same method. The remaining studs will be the same distance apart as the first two, so that you can simply measure to find them. Use a ¾-inch spade bit to drill a hole through the center of each stud. In a plaster wall, cut the rectangular hole with a chisel and then saw through the underlying lath (*page 68*).

2 Fishing the cable. Remove a knockout tab in the bottom of the existing box and push a fish tape through the hole and down into the wall. Hook this tape with a second tape pushed through the hole in the nearest stud. With the second tape, pull the first through the stud. Now release the second tape, attach cable to the free end of the first tape, and pull tape and cable back through the stud and into the existing box. Using the same two-tape method, fish the cable through the adjoining studs and into the opening you have cut for the new box. Clamp the cable to both boxes (*pages 84-85*), install the new box (*pages 64-69*) and patch up the wall (*pages 104-105*).

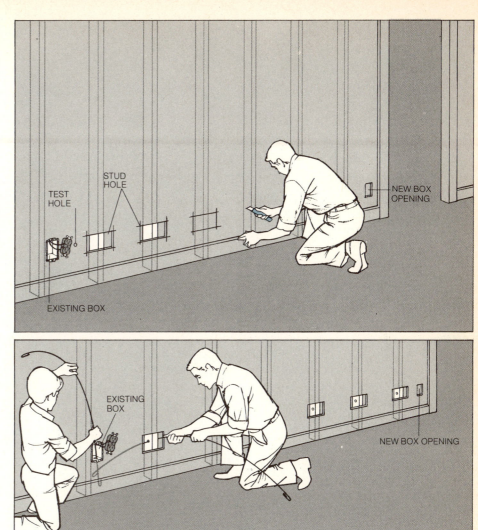

Concealing Cable behind a Baseboard

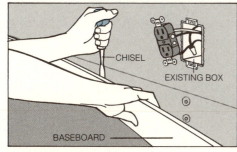

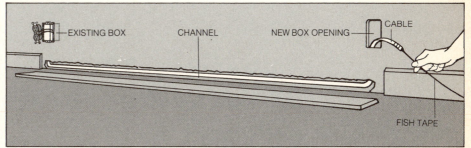

1 Removing the baseboard. Make an opening for the new box (*pages 64-69*). Directly below the existing box, insert the blade of a wood chisel between the wall and the top of the baseboard, and tap the chisel gently with a mallet until the baseboard begins to separate from the wall. Move the chisel a few inches along the baseboard and repeat the process until you have loosened an entire section. Use the chisel as a lever with one hand and pull the section from the wall with the other hand.

2 Cutting a channel. For a plaster wall, chisel a 1-inch hole through the plaster and lath below the existing box in the section behind the baseboard. Make an opening for the new box (*page 66*) and break an identical hole below it. With a hammer and cold chisel, cut a channel between the two holes. Fish cable from the existing box to the first hole, run it along the channel and fish it to the new box opening. Clamp the cable to both boxes and in-

stall the new box. (Although electrical codes stipulate that cable run in notches be covered by 1/16-inch metal plate—to protect the cable from being punctured by nails—this requirement is widely disregarded.) In plasterboard, cut a strip from the section behind the baseboard. Drill holes through the exposed studs, fish the cable into the existing box, run it through the studs and then fish it to the new box opening.

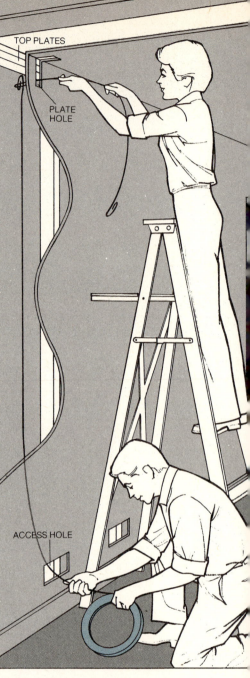

Taking Cable from Wall Box to Ceiling Box

1 Making access holes. Cut an opening for the new ceiling box between two joists (*pages 70-75*). Follow along these joists—not across—to the wall and mark the point midway between them at the top of the wall. If the existing wall box from which you plan to run new cable does not lie directly below this point, you will have to make access holes in the wall to bring cable horizontally to position (*page 79, Steps 1 and 2, top*). From the mark made at the corner of the wall and ceiling directly above the box or hole, outline a 2-inch-long rectangle on the ceiling and a 4-inch-long rectangle on the wall.

Cut the plasterboard or score the plaster along the rectangles with a utility knife, and cut through ceiling and wall along the outlines. In plasterboard, break into the ceiling rectangle with a cold chisel and a ball-peen hammer, and cut away the broken fragments with the utility knife. Chisel out the wall rectangle. In plaster, chisel out both rectangles down to the lath, then cut away exposed wood lath with a keyhole saw, or metal lath with a saber saw.

2 Fishing cable up to ceiling level. Insert one end of a fish tape through the wall access hole beneath the ceiling rectangle or, if the existing box lies directly below the ceiling opening, through a knockout hole in the box. Have a helper push the fish tape upward to the top plates. Drill through any fire blocking as shown on page 79, Steps 1 and 2, top. With a second fish tape, hook the top of the vertical tape and pull it out. Disconnect the two tapes, attach the cable to the top end of the vertical tape (*page 77, Step 3*), and have your helper draw the cable down behind the wall and out through the wall access hole (or knockout hole).

3 **Fishing cable above the ceiling.** Have your helper insert one end of a fish tape into the ceiling box opening and push the tape between the joists past the rectangular ceiling hole at the top plates. Insert a second fish tape through this ceiling plate hole and hook it over the first tape; then hold it down firmly as your helper slowly pulls his tape back until the two tape ends hook together. Reversing direction, pull your tape completely out of the ceiling plate hole. Disconnect the two tapes, attach the end of the cable to the upper tape and feed the cable into the ceiling plate hole as your helper slowly draws the cable toward and out of the ceiling box opening.

4 **Fastening the cable to the top plates.** If the wall is less than ½ inch thick, you must cut a channel for the cable in the top plates. Invert the handle of a keyhole saw and make two parallel vertical cuts, ½ inch deep and ¾ inch apart, across the exposed top plates. Chisel out the wood between the cuts. Staple the cable in the groove just below the top edge of the plates.

If the wall is more than ½ inch thick, you need not chisel a groove; simply staple the cable to the top plates. In either case, complete the job by patching the two access holes with plasterboard or plaster (*pages 104-105*).

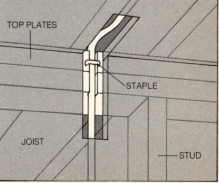

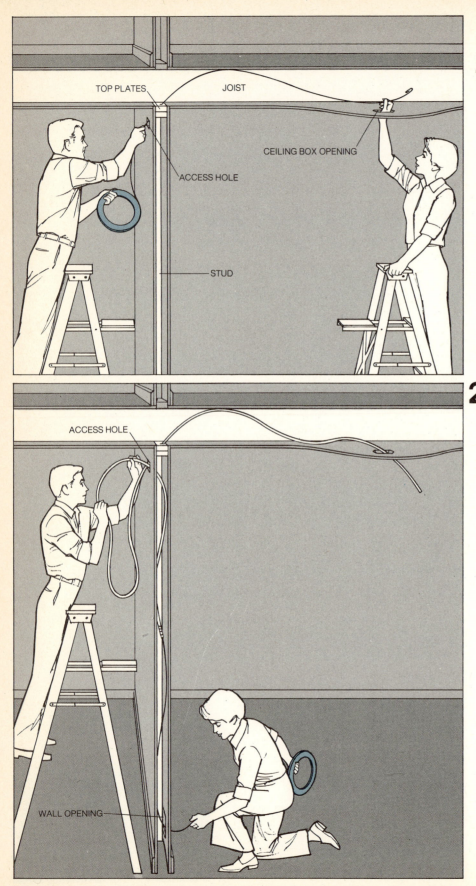

TOP PLATES · JOIST

CEILING BOX OPENING

ACCESS HOLE

STUD

ACCESS HOLE

WALL OPENING

Working from an Adjoining Room

1 **Fishing cable from an adjoining room.** In some cases, this method may be simpler than the one on pages 80-81. Begin in the same way: Cut a ceiling box opening and mark the midpoint between joists at the top of a wall (*page 80, Step 1*). Then locate the equivalent point on the other side of that wall by measuring off the distance from a corner or a doorway. In the adjoining room, drill a ⅛-inch hole 4 inches below the mark at the ceiling. Insert a wire to locate the top plates, and make an access hole 2 inches high and 1 inch wide (*page 80, Step 1*) just below the plates.

Using a drill with an extension and a ¾-inch bit, and starting about ½ inch inside the plate edge, bore a hole through the plates as close to the vertical as possible. Push a fish tape through this hole into the ceiling space to the ceiling box opening. Have a helper pull the tape through the opening and attach cable to it (*page 77, Step 3*). Then, reversing direction, pull the cable through the plates and into the adjoining room.

2 **Fishing cable to the wall box.** Have your helper push a fish tape to the top of the wall through the wall opening in the room being wired. Insert a second tape through the access hole in the adjoining room, hook the first one and pull it into the adjoining room. Attach the cable to the first tape (*page 77, Step 3*), and have your helper pull it through the wall box opening. Then patch up the access hole (*pages 104-105*).

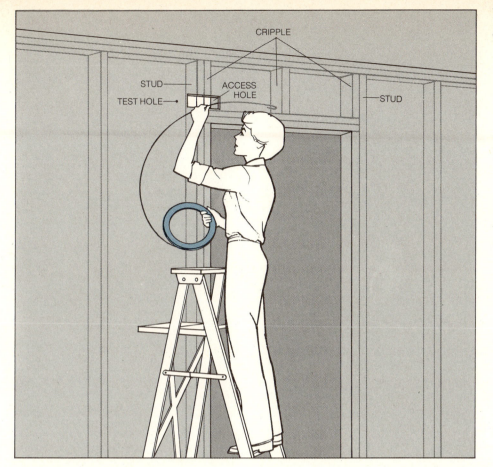

The Best Way to Get around a Doorway

Fishing cable through the cripples. Bore a
⅛-inch hole through the wall about 6 inches above
and outside the doorway, angling the drill toward
the doorway. Insert a piece of stiff wire to locate
the exact position of the stud next to the door
and cut an access hole (*page 79, Step 1*) wide
enough to expose both the stud and the first cripple—a short vertical support—over the doorway.
Use the fish tape to locate the next cripple, cut
an access hole to expose it and repeat the process until you have cut an access hole at the
stud on the other side of the doorway. Drill holes
through the studs and cripples (*page 79, Step 1*)
and fish the cable through the holes and over the
doorway. Patch up the holes (*pages 104-105*).

Two Boxes Back to Back

Installing outlet boxes back to back. Measure
the distance from a reference point such as a
doorway to the existing outlet box, and mark the
equivalent point on the other side of the wall.
Turn off the current, remove the cover plate from
the existing box and poke a coat-hanger wire into
the space between the sides of the box and the
edges of the opening to find the stud to which
the box is attached. Go to the adjoining room
and mark a second point about 8 inches from the
first, measuring away from the stud supporting
the existing box. Cut an opening for the new outlet box at the second point (*pages 64-69*).

Have a helper push a fish tape into the wall
through a knockout hole in the existing box,
and then hook this tape with a second tape
inserted through the new opening. Pull the first
tape through the new opening, attach 3 feet
of cable to it, and have your helper pull the cable
back through the wall space and into the existing
box. Attach the cable to both the existing and
new box, and install the new box (*pages 64-69*).

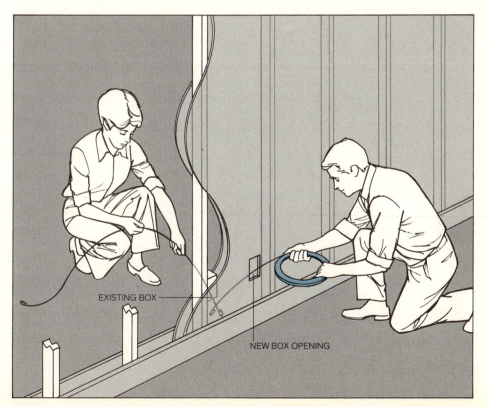

Connecting Cable to Boxes

Cable enters a box through one of its "knockout holes," small sections on the box that contain removable disks, or knockouts. Choose the hole that provides the most direct path for the cable. Remove the knockout from this hole and fasten the cable to the box. There are two types of knockout holes, and cable is connected to each in a different way.

One type of knockout is U shaped, has a slot in the middle and is almost exclu-sively found in wall boxes. Cable entering this hole is fastened with an internal clamp (below), usually supplied with the box, that accepts armored or plastic-sheathed cable, but not conduit.

Round knockout holes are provided in ceiling and junction boxes as well as in wall boxes, and they can accept conduit as well as cable. The cable is fastened at the hole by a two-part connector (opposite), which consists of a clamp that holds the cable and a threaded tube that fits into the knockout hole. However, these connectors are purchased separately.

Usually, you will connect cable to a new box as shown in Steps 1, 2 and 3 below. If you are adding cable to a pre-viously installed wall box, however, start opening the U-shaped knockout from within the box with a nail set and hammer, then insert the looped end of a coat-hanger wire between the wall and the box and pull the knockout back against the box. Fish cable into the box through the knockout hole and connect the cable with an internal clamp. To connect cable with a two-part connector to a previously installed ceiling or junction box, you must remove a section of the ceiling or wall to get at the outside of the box (pages 64-75). Then follow the procedures shown here for new work.

Attaching an Internal Clamp

1 **Removing the U-shaped knockout.** Insert the tip of a screwdriver into the slot of the knockout and pry the knockout away from the box. Work the knockout back and forth until it breaks free.

2 **Adapting clamp.** Although some internal clamps are designed exclusively for plastic-sheathed cable, the most common type, shown below, has extra metal loops to accommodate armored cable. To prepare the latter type of clamp for plastic-sheathed cable, cut off the metal strips holding the loops with metal snips.

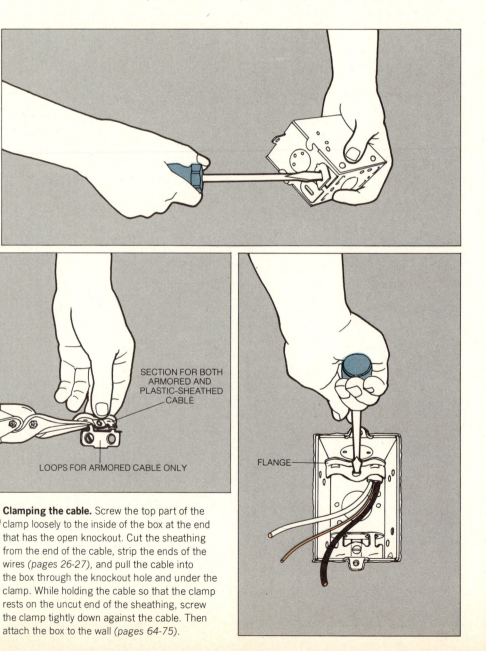

SECTION FOR BOTH ARMORED AND PLASTIC-SHEATHED CABLE

LOOPS FOR ARMORED CABLE ONLY

FLANGE

3 **Clamping the cable.** Screw the top part of the clamp loosely to the inside of the box at the end that has the open knockout. Cut the sheathing from the end of the cable, strip the ends of the wires (pages 26-27), and pull the cable into the box through the knockout hole and under the clamp. While holding the cable so that the clamp rests on the uncut end of the sheathing, screw the clamp tightly down against the cable. Then attach the box to the wall (pages 64-75).

Attaching a Two-Part Clamp

1 **Removing the round knockout.** Prop the box on a firm surface and open the knockout hole with a nail set and a hammer. Then grip the knockout with a pair of pliers and work it free.

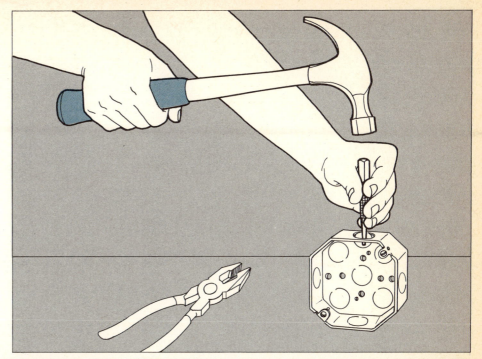

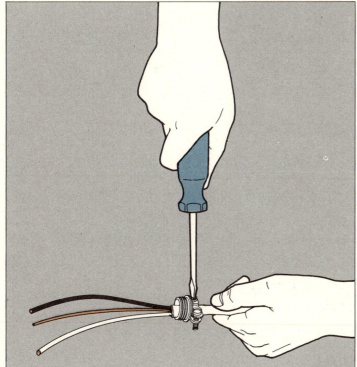

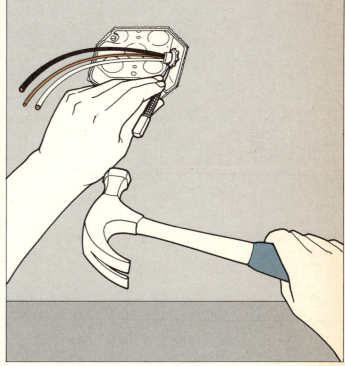

2 **Fastening the connector to the cable.** Cut the sheathing from the end of the cable and strip the end of the wires (*pages 26-27*). Slip a connector for plastic-sheathed cable onto the sheathing, with the threaded end facing the stripped wires and flush to the end of the sheathing. Then tighten the clamp screws onto the cable.

3 **Fastening the connector to the box.** Insert the stripped wires and the threaded end of the connector into the box through the knockout hole. Slip the connector nut over the wires and screw the nut onto the connector. Then attach the box to the wall or ceiling (*pages 64-75*). To tighten the connector nut, position a nail set against one of the points that protrudes from the nut and hammer the nail set to turn the nut.

Installing Switch-controlled Light Fixtures

Once you know how to install boxes and run cables (pages 64-83), installing new circuits for lights and switches becomes simple, since the wiring connections are similar to those that are made in replacing old ones (pages 40-51). You can install special-purpose lights in a living room, a new fixture in a bedroom, or a combination of switch, light and receptacle for extra light over a kitchen work area and additional outlets for appliances. Using two switches, you can control lights from different locations in a stairwell, garage or two-entrance room.

In every case you must run cable from an existing power source (page 63) to newly installed switch and fixture outlet boxes. The route of the cable depends on the location of the circuit you will tap for power and the layout of your home. The route determines the order in which power moves to each device in the circuit. That order, in turn, determines how many wires the cables contain and the way those wires must be joined.

The drawings on these pages show how these relationships work in practice. When a power source is available in an unfinished attic (opposite, top), the easiest cable route for a ceiling light in a room beneath runs through the attic to the new light and then down through the wall to a single-pole switch. If the attic is inaccessible but power is available at a baseboard receptacle (below), the cable will run through the wall to the switch, then up and along the ceiling to the light. If there is a power source available in an adjacent hallway (opposite, bottom), the cable might run across and up to a switch, then to a ceiling light and—if you install a three-conductor cable between the switch and the light —power can be continued on to supply a wall receptacle.

A new middle-of-the-run switch. The cable for this new circuit starts at a power source, then runs to the switch and on to the light. In this arrangement, the switch wiring (inset, bottom left) is called middle-of-the-run and the light-fixture wiring (inset, right) end-of-the-run. Throughout, two-conductor cable with ground wire is used.

At the switch, the hot black wire of the incoming cable is connected to one terminal and the black outgoing wire is connected to the other terminal. The white neutral wires, which must provide an unbroken path for current from the light back to the service panel (pages 12-13), are joined with a wire cap. The ground wires, which must also be uninterrupted, are joined to each other, to the metal box and to the green grounding screw on the strap of the switch. At the light fixture, the hot wire from the switch is connected to the fixture's black wire, the neutral wire to the fixture's white wire and the bare ground wire to the box.

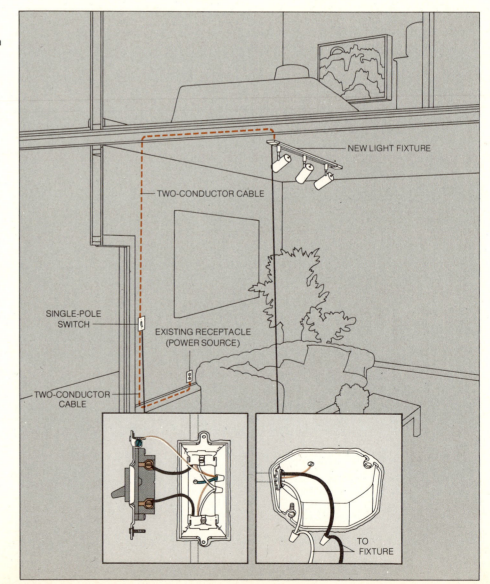

NEW LIGHT FIXTURE

TWO-CONDUCTOR CABLE

SINGLE-POLE SWITCH

EXISTING RECEPTACLE (POWER SOURCE)

TWO-CONDUCTOR CABLE

TO FIXTURE

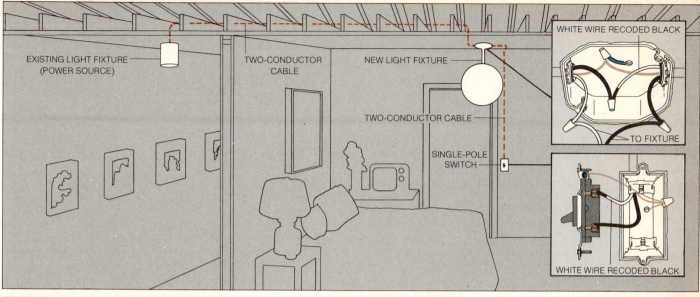

EXISTING LIGHT FIXTURE (POWER SOURCE)

TWO-CONDUCTOR CABLE

NEW LIGHT FIXTURE

TWO-CONDUCTOR CABLE

SINGLE-POLE SWITCH

WHITE WIRE RECODED BLACK

TO FIXTURE

WHITE WIRE RECODED BLACK

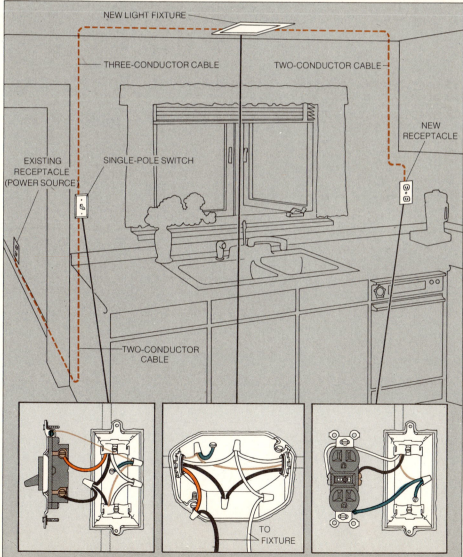

NEW LIGHT FIXTURE

THREE-CONDUCTOR CABLE

TWO-CONDUCTOR CABLE

NEW RECEPTACLE

EXISTING RECEPTACLE (POWER SOURCE)

SINGLE-POLE SWITCH

TWO-CONDUCTOR CABLE

TO FIXTURE

A switch loop. With cable running first to the light fixture, then to the switch, the fixture wiring *(inset, top)* is called middle-of-the-run and the switch wiring is a switch loop *(inset, bottom).* Two-conductor cable with ground is used throughout.

At the new fixture, the incoming white neutral wire is joined to the white fixture wire. The black wire of the incoming cable is joined to the white outgoing wire, to power the switch; the black outgoing wire, bringing power from the switch, is connected to the black fixture wire. At the switch, the white wire is fastened to one terminal, the black to the other, forming the loop that carries power to the switch and back to the light. Be sure to indicate that this white wire is hot by recoding it black at the switch and ceiling box. Pigtail the cable's bare ground wire and run one wire to the metal box, another to the grounding terminal on the switch.

A combination circuit. This circuit runs from an existing receptacle to a light and an unswitched receptacle. To provide a switched hot wire for the light and an unswitched one for the receptacle, a three-conductor cable runs from the switch to the light.

At the switch, the incoming neutral and ground wires are connected as shown opposite. The black incoming wire is connected with a jumper to the switch and the black wire of the three-conductor cable; it runs directly to the receptacle. The cable's red wire carries power from the switch to the black fixture wire; the white wire of the cable is joined to the white fixture wire and the white wire running to the receptacle. The receptacle wiring is end-of-the-run *(page 55)*, and the receptacle has a ground-fault interrupter because it is within 6 feet of the sink.

Wiring New Three-Way Switches

A pair of three-way switches, controlling a light from two locations, is more than a frill. Stairways should have switches at top and bottom. Lights in long halls are best controlled from either end, and kitchen and dining-room lights gain in convenience when they can be switched from two entrances. And you will never have to enter a dark garage or leave a light burning if there are switches in the house and in the garage.

How three-way switches are installed depends on the relationships of the existing circuit to be tapped for the power source, the switches themselves and the light they control. Three common arrangements are shown here—in a stairway, a dining room and a garage. The variation you choose will depend on the construction of your house and on a power source for the new wiring.

If a power source is available in an attic, you can run cable to the light fixture and then on to the switches (below). Taking power from a basement, you may run cable first to the switches and then to the fixture (opposite, top). When tapping a baseboard outlet, the wiring arrangement at the bottom of the opposite page is often the most convenient.

Three-conductor grounded cable—black and red hot wires plus a white neutral wire and a bare ground wire—is required for part of a three-way circuit. (The arrangement of circuits determines which part.) White wires sometimes serve as hot wires in a three-way switch loop; be sure to recode them black, so they cannot be mistaken for neutral wires. The terminals in three-way switches vary; make the tests shown on page 46 to label the terminals. When running three-conductor cable, you often need more wires in an outlet box than the National Electrical Code permits (page 64); if necessary, install a larger box or gang two boxes together.

From light to switch to switch. For the stairway below, with three-way switches, a two-conductor cable runs from the power source to the light fixture. Another two-conductor cable runs from the fixture to the upstairs switch, and a three-conductor cable runs between the two switches. The only neutral white wire is the one in the cable from the power source; all other white wires are part of a switch loop and therefore are coded black.

At the fixture, connect the incoming white wire to the fixture, the incoming black wire to the white wire of the outgoing cable, and the outgoing black wire to the fixture. At the upstairs switch, connect the incoming white wire to the black wire of the three-conductor cable, the incoming black wire to the common terminal, and the outgoing red and white wires to the traveler terminals (page 46). Downstairs, connect the black wire to the common terminal, the red and white wires to the traveler terminals. Connect all ground wires to each other, to the outlet boxes and to the switches.

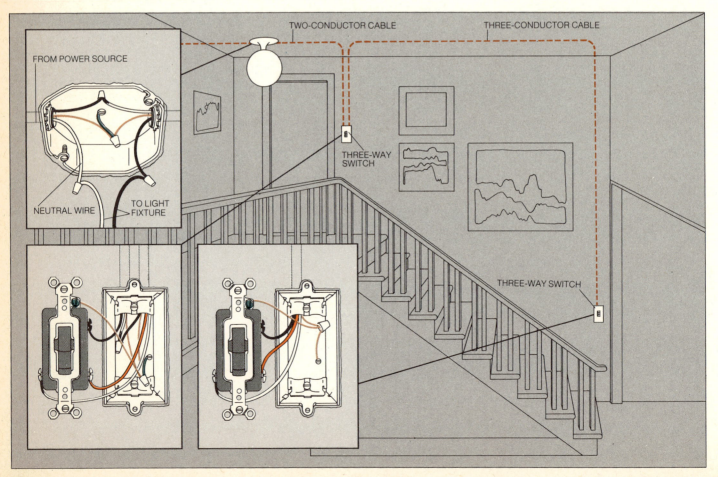

FROM POWER SOURCE

NEUTRAL WIRE

TO LIGHT FIXTURE

TWO-CONDUCTOR CABLE

THREE-CONDUCTOR CABLE

THREE-WAY SWITCH

THREE-WAY SWITCH

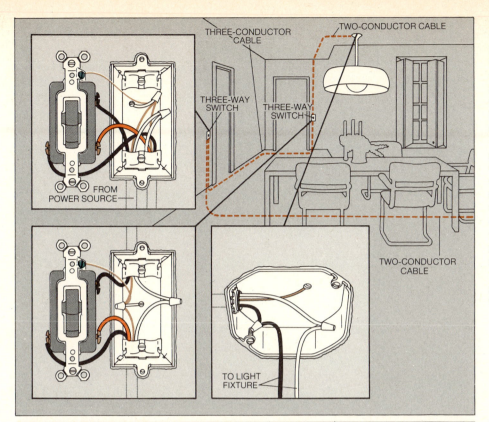

THREE-CONDUCTOR CABLE

TWO-CONDUCTOR CABLE

THREE-WAY SWITCH

THREE-WAY SWITCH

FROM POWER SOURCE

TWO-CONDUCTOR CABLE

TO LIGHT FIXTURE

From switch to switch to light. From a power source in the basement under this dining room, a two-conductor cable runs to the first three-way switch *(left)*, a three-conductor cable runs from the first to the second switch, and a two-conductor cable from the second switch to the overhead light fixture. In this wiring arrangement, the white wire in all three cables is neutral.

At the first switch, join the two white wires with a wire cap; connect the incoming black wire to the common terminal of the switch, and the outgoing black and red wires to the traveler terminals. At the second switch, connect the incoming red and black wires to the traveler terminals, join the white wires and fasten the outgoing black wire to the common terminal. At the fixture, connect the black and white wires to the light. Join all ground wires to each other, to the boxes and to the switches.

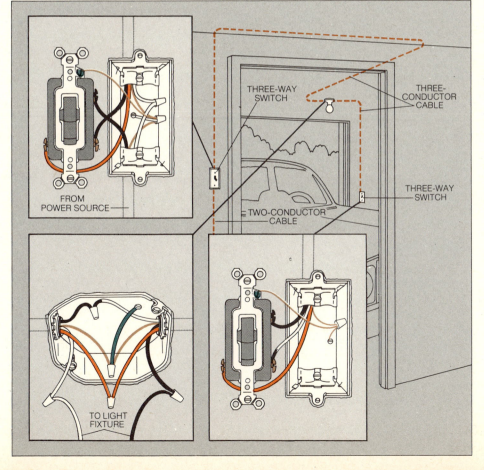

THREE-WAY SWITCH

THREE-CONDUCTOR CABLE

THREE-WAY SWITCH

FROM POWER SOURCE

TWO-CONDUCTOR CABLE

TO LIGHT FIXTURE

From switch to light to switch. In a circuit extended from a house to a garage, a two-conductor cable runs from the power source to the house switch, a three-conductor cable from the house switch to the light fixture, and a three-conductor cable from the fixture to the garage switch. The white wire running from the power source to the light fixture is neutral; beyond the fixture, the white wire is used as a hot wire in a switch loop and must be recorded black.

At the house switch, connect the incoming black wire to the common terminal, join the two neutral wires with a wire cap and fasten the black and red wires of the outgoing cable to the traveler terminals. At the fixture, connect the incoming white wire to the fixture, join the two red wires, join the incoming black wire to the outgoing white wire and connect the outgoing black wire to the fixture. At the garage switch, fasten the black wire to the common terminal, the red and white wires to the traveler terminals. Connect all of the ground wires to each other and to the boxes and switches.

Dramatic Effects with Built-in Lighting

Once you have learned the techniques for installing a new fixture and switch, you can use this know-how to build in lighting systems that become unobtrusive permanent parts of a room and provide exceptionally comfortable, even illumination. Such "structural lighting"—cornices *(below)* and valances, soffits, coves, luminous ceilings and track lighting *(following pages)*—is expensive when installed by professionals but it requires only a knowledge of simple carpentry and basic wiring techniques. Many such devices are simply rows of fluorescent

fixtures concealed behind a board to direct light to the desired area.

Fluorescent fixtures are used because they produce an even, diffused light and come in various lengths that can be joined with asbestos-insulated wire and a device called a channel connector to adapt to almost any size installation. Use rapid-start fixtures *(page 38)* but be sure they are properly grounded. Install a switch and cable to operate the lights *(page 76)*, bringing cable to a point near the mounting block location and leaving 6 inches of slack so that the cable can

reach the nearest fluorescent channel. No box is needed, since the fluorescent channel serves that purpose.

Because of the size and weight of the fixtures and materials used for built-in lighting, the parts must be securely attached to the ceiling or wall with the correct fasteners. Use wood screws to attach components to wood beams or studs; for plaster, wallboard or brick use screws with anchors. Paint the inner surfaces of all installations flat white to minimize glare and reflect the maximum amount of light.

Building a Cornice Light

Fitting the boards. The cornice is a board in front of a row of fluorescent lights that are attached to a mounting board on the ceiling about 6 inches from a wall *(above)*. Because illumination is directed downward—over the stone section in the sketch above—the installation makes the ceiling appear higher than it actually is. To mount the

lights, you need a 1-inch-thick strip of wood the width of the fluorescent channel and long enough to hold the number of lights you plan to install. The cornice board, which is screwed to the mounting block, should measure at least 6 inches wide, ¾ inch thick and should be as long as the wall you are planning to light.

1 Preparing to mount the lights. Cut the mounting block and attach asbestos sheeting to its surface to protect the wood from the heat of the ballasts. Punch out the tabs at both ends of all the fluorescent channels except the tab at the far end of the last channel in the row.

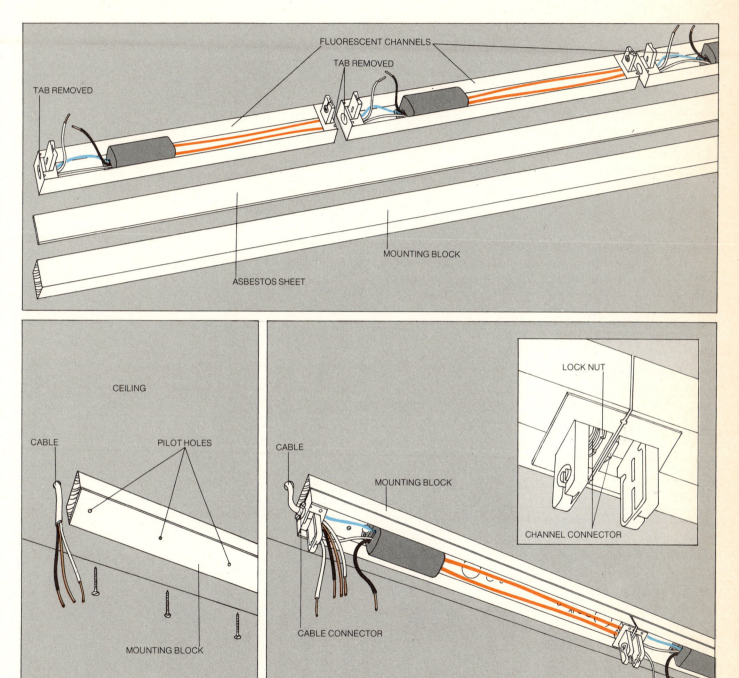

FLUORESCENT CHANNELS

TAB REMOVED

TAB REMOVED

MOUNTING BLOCK

ASBESTOS SHEET

CEILING

CABLE

PILOT HOLES

MOUNTING BLOCK

CABLE

MOUNTING BLOCK

CABLE CONNECTOR

LOCK NUT

CHANNEL CONNECTOR

2 Mounting the block. Drill pilot holes for screws in the mounting block at 4-inch intervals. Draw a line on the ceiling that is parallel to the wall and 6 inches away from it. With an assistant, hold the mounting block inside the line and mark the ceiling for screws. Attach the block to the ceiling.

3 Mounting the lights. Screw the fluorescent channels to the block with their ends butted. Secure the cable with two-part cable connector in the hole in the end of the first channel on the block. Join the channels with channel connectors (inset). Slip the connector stud through the holes in the channel ends and secure with a lock nut.

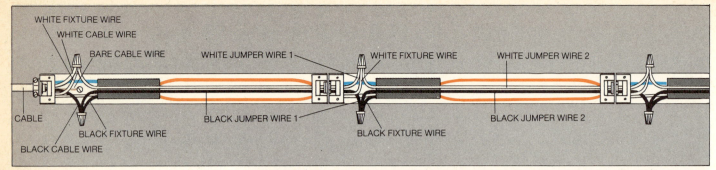

WHITE FIXTURE WIRE
WHITE CABLE WIRE
BARE CABLE WIRE
WHITE JUMPER WIRE 1
WHITE FIXTURE WIRE
WHITE JUMPER WIRE 2
CABLE
BLACK FIXTURE WIRE
BLACK CABLE WIRE
BLACK JUMPER WIRE 1
BLACK FIXTURE WIRE
BLACK JUMPER WIRE 2

4 **Wiring the lights.** Connect the black cable wire to the black wire in the first fixture and to a black jumper long enough to reach the black wire in the next fixture. In the same manner, connect the white cable wire, the white fixture wire and a white jumper. Attach the bare cable wire to the channel with a screw. Run both jumper wires through the channel connector to the next fixture and attach them to the wires in that fixture. Join additional fixtures the same way.

5 **Attaching the cornice board.** Drill pilot holes at 4-inch intervals ½ inch below the top of the cornice board. With a helper holding one end of the board, place it flat against the outer edge of the mounting block with its top against the ceiling. Mark the mounting block for matching pilot holes, drill them and screw on the cornice board.

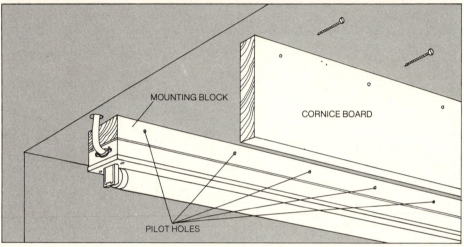

MOUNTING BLOCK

CORNICE BOARD

PILOT HOLES

Valance, Cove and Soffit Lighting

Valance lighting. The construction of a valance light (*right*) is basically the same as that of a cornice light, but the fixtures are attached to a wall at least 10 inches below the ceiling. The valance is usually installed above a window and projects light up onto the ceiling and down onto the draperies. Align the fixtures end to end on a mounting block and wire them to a cable in the same manner as for cornice lighting. The valance board, with end boards to conceal the row of fixtures, is attached to the mounting block with wood screws and angle irons as shown in the inset.

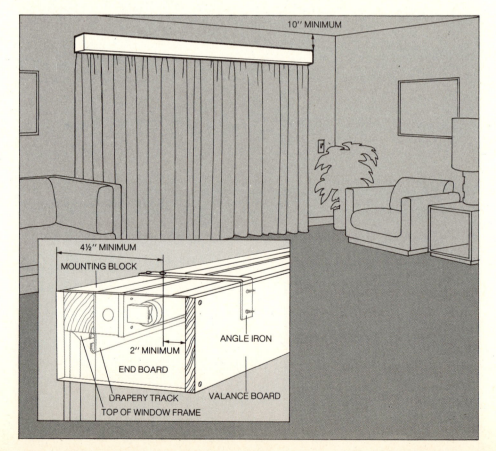

10" MINIMUM

4½" MINIMUM
MOUNTING BLOCK
2" MINIMUM
ANGLE IRON
END BOARD
DRAPERY TRACK
TOP OF WINDOW FRAME
VALANCE BOARD

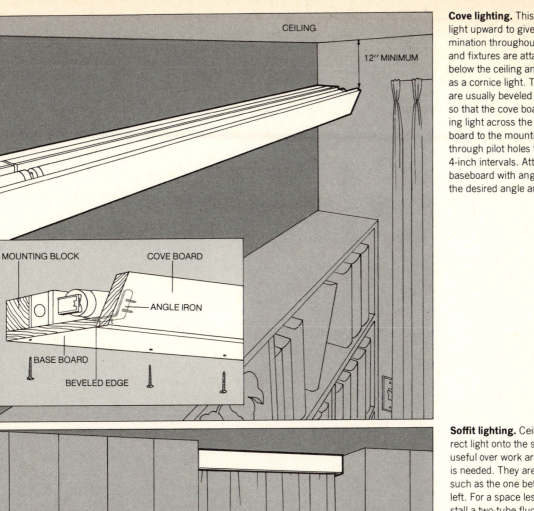

CEILING

12" MINIMUM

MOUNTING BLOCK   COVE BOARD

ANGLE IRON

BASE BOARD

BEVELED EDGE

Cove lighting. This type of installation directs all light upward to give very diffuse reflected illumination throughout a room. The mounting block and fixtures are attached to the wall about 1 foot below the ceiling and wired in the same manner as a cornice light. The baseboard and cove board are usually beveled on their joining edges *(inset)* so that the cove board will angle outward, spreading light across the ceiling. Secure the baseboard to the mounting block with wood screws through pilot holes that have been drilled at 4-inch intervals. Attach the cove board to the baseboard with angle irons that are bent to the desired angle and are spaced 4 inches apart.

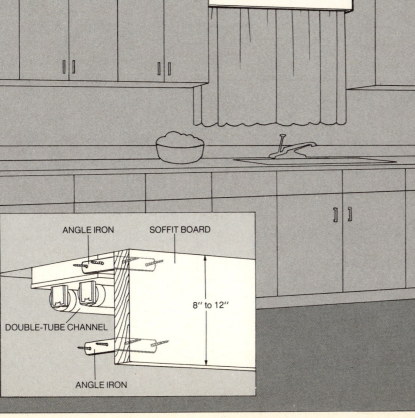

ANGLE IRON   SOFFIT BOARD

8" to 12"

DOUBLE-TUBE CHANNEL

ANGLE IRON

Soffit lighting. Ceiling-mounted soffit lights direct light onto the surface directly below, and are useful over work areas where strong illumination is needed. They are usually built into a recess such as the one between the kitchen cabinets at left. For a space less than 18 inches deep, install a two-tube fluorescent light; for a recess from 18 to 24 inches deep install a three-tube fixture. Attach the channel directly to the ceiling without a mounting block. The soffit board, 8 to 12 inches wide, should be attached to the sides of the cabinets with angle irons *(inset)*. Connect the lights to a cable in the same way as for cornice, valance or cove lighting.

Making a Luminous Ceiling

A luminous ceiling—plastic panels in an aluminum framework of cross Ts and runners hung beneath ceiling-mounted fluorescent fixtures—can provide unobtrusive, even illumination. It can also camouflage such problems as piping, damaged surfaces or a too-high ceiling.

The parts required are available in kit form or they can be bought separately. The frame comes in sections that can be cut to the desired length or joined together with metal splicers. If you are buying the fluorescent lights separately, the best choice is usually the 4-foot,

40-watt rapid-start type. To estimate the number you need, sketch a plan like the one below. The fixtures are generally mounted in parallel rows about 2 feet apart, with the outermost rows some 8 inches from the side walls, and the row ends some 8 inches from the end walls. A room 8 by 10 feet needs six 4-foot fixtures. If the ceiling does not already have a box controlled by a switch, run the necessary cable, install a box and switch. Paint all surfaces above the luminous ceiling with two coats of flat white paint to reflect light downward.

1 Mounting the lights. Draw lines on the ceiling about 2 feet apart for the rows of lights. Remove the knockout tabs from the ends of the lights that butt together. Using the screw holes in the channels as a guide, install screw anchors along the lines and screw the channels to the ceiling, butting the ends. Connect the ends with channel connectors (*page 91, Step 3*).

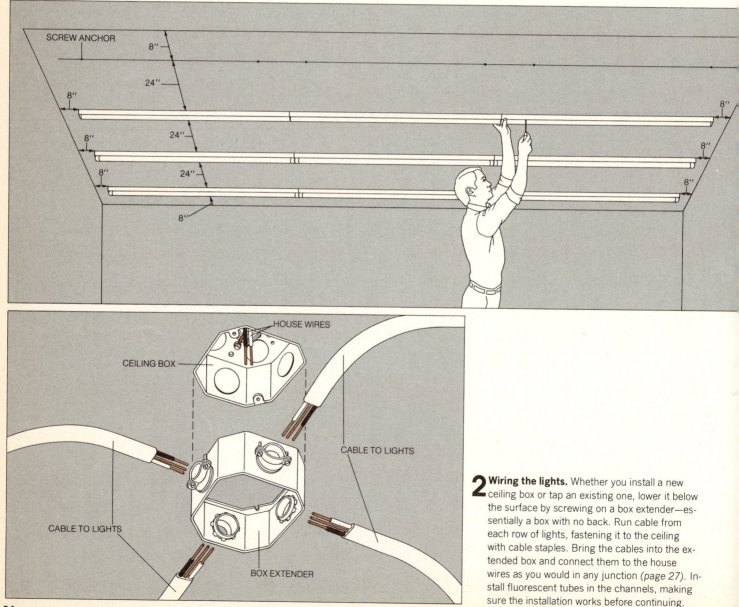

SCREW ANCHOR
8″
24″
8″
8″
24″
8″
8″
24″
8″
8″
8″
8″

HOUSE WIRES
CEILING BOX
CABLE TO LIGHTS
CABLE TO LIGHTS
BOX EXTENDER

2 Wiring the lights. Whether you install a new ceiling box or tap an existing one, lower it below the surface by screwing on a box extender—essentially a box with no back. Run cable from each row of lights, fastening it to the ceiling with cable staples. Bring the cables into the extended box and connect them to the house wires as you would in any junction (*page 27*). Install fluorescent tubes in the channels, making sure the installation works before continuing.

3 Attaching the edge framing. Draw a horizontal line around the walls of the room at the desired height for the new ceiling (which should be no lower than 7½ feet). With a carpenter's level, make sure the line is perfectly horizontal. Nail the edge framing—aluminum strips with an L cross section—to the wall, using the holes provided.

4 Installing the main runners. To position the first support runner of the aluminum frame, measure out from a side wall the distance specified in the manufacturer's directions, usually between 2 and 4 feet. At that distance, draw a line along the existing ceiling and parallel to the side wall.

Install eye screws at 2-foot intervals along the drawn line. Insert hanger wires, secure them by twisting and bend the free ends to a 90° angle. Then set the main runner in place, with its ends resting on the edge framing. Thread each hanger wire through one of the holes provided in the runner. Check to be sure the runner is level, then secure each wire by twisting it back around itself. The other main runners are installed the same way, at intervals determined by the size of the plastic panels that are to be used.

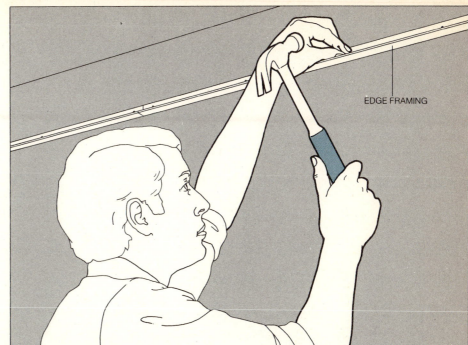

EDGE FRAMING

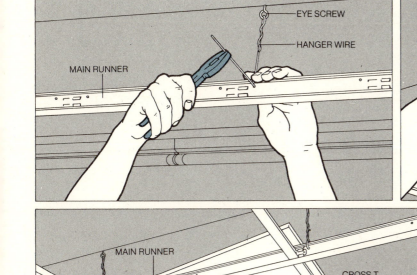

EYE SCREW

HANGER WIRE

MAIN RUNNER

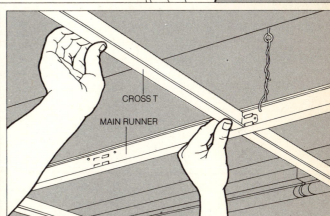

CROSS T

MAIN RUNNER

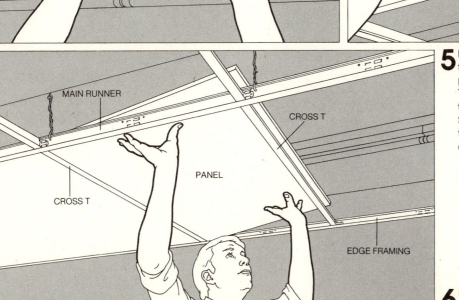

MAIN RUNNER

CROSS T

PANEL

CROSS T

EDGE FRAMING

5 Positioning the cross Ts. Space cross Ts across the runners at intervals equal to the width of the plastic panels. Place one end of the first cross T in each row on the edge framing, then snap the other end into the slot provided in the runner. Snap the other cross Ts in that row between the intervening runners and install the Ts of the remaining rows, using the same procedure.

6 Installing the panels. Lift each of the panels diagonally up through the framework, turn it to the horizontal and rest its edges on the flanges of the cross Ts and main runners or edge framing.

Setting Up Track Lighting

Track lighting systems, developed for museum and store displays, are now used to highlight paintings, provide reading lights and concentrate illumination on work areas. The track is an elongated receptacle, with current-supplying contacts in long grooves. The lights plug into the grooves and can be moved to any position and aimed in any direction.

The track, usually 4 or 8 feet long, is prewired, but it must be connected to the house wiring. Sections snap together with connectors—straight, T-shaped, X-shaped or right angled—to create arrangements of many designs. Installation methods vary, and units from one manufacturer may not fit those of another.

Before you buy a track system, first select a position for the track and decide where you will connect it to the house wiring. The light should strike a wall at about a 30° to 45° angle. On an 8-foot ceiling, place the track 2 to 2½ feet from the wall; on a 9-foot ceiling, about 3½ feet. In most rooms, a ceiling box is not close enough to be used as a power source. The alternatives are to plug the track into a receptacle with a special adapter *(bottom)* or to install a new ceiling box closer to the wall *(page 64)*.

Versatile Fittings

Track lighting systems can be used for many more purposes than the standard strip of spotlights that is shown in the drawing at near right. You can insert adapters into the track to substitute fluorescent fixtures or even a chandelier *(center)*. You can also clip in an adapter that will provide you with a grounded receptacle useful in work areas. A spotlight can also be clipped into a special adapter *(far right)* that will turn the spotlight into a plug-in lamp that can be located near an outlet anywhere in the room.

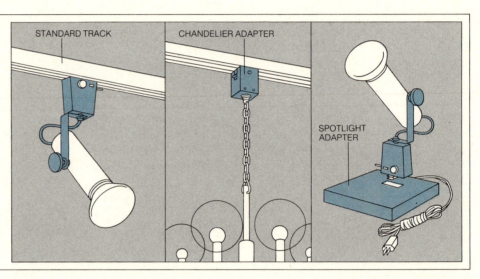

STANDARD TRACK

CHANDELIER ADAPTER

SPOTLIGHT ADAPTER

A Plug-in Track

Attaching the track and adapter. The simplest way to install track lighting is to use an adapter fitted with a power cord and plug like that on a lamp (you may have to find a way to conceal the adapter's cord behind curtains or a bookcase). Mark a line on the wall or ceiling where you want to place the track. Hold the track on the line and mark the locations of the screw holes. Then drill pilot holes and attach the track with screws, using expansion anchors if necessary.

To avoid any chance that an attaching screw might conduct electricity onto the metal track itself, use an insulated washer big enough to keep the head of each screw from touching the track. When the track is in place, push the adapter onto the end of the track so that the track wires enter the adapter connectors. Snap on the adapter cover and insert the power-cord plug in a receptacle.

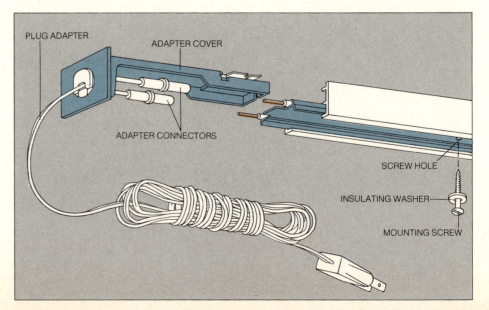

PLUG ADAPTER

ADAPTER COVER

ADAPTER CONNECTORS

SCREW HOLE

INSULATING WASHER

MOUNTING SCREW

Wiring a Track to a Box

1 Bringing power to the track. Most tracks are connected to a ceiling box by three parts: an electrical connector that fits into a track connector, which in turn is held by a box adapter screwed to the box. To position the track, first screw the track connector to the box adapter. Push the wires of the electrical connector through the slot in this two-piece mounting assembly and then attach the wires to those in the box—black to black and white to white. Attach the mounting assembly to the box tabs.

2 Placing the track. Hold a ruler flat against the wall or ceiling with the edge of the ruler lined up with the center slot in the track connector (below). Draw a line, moving the ruler along as necessary, to the point where the track will end.

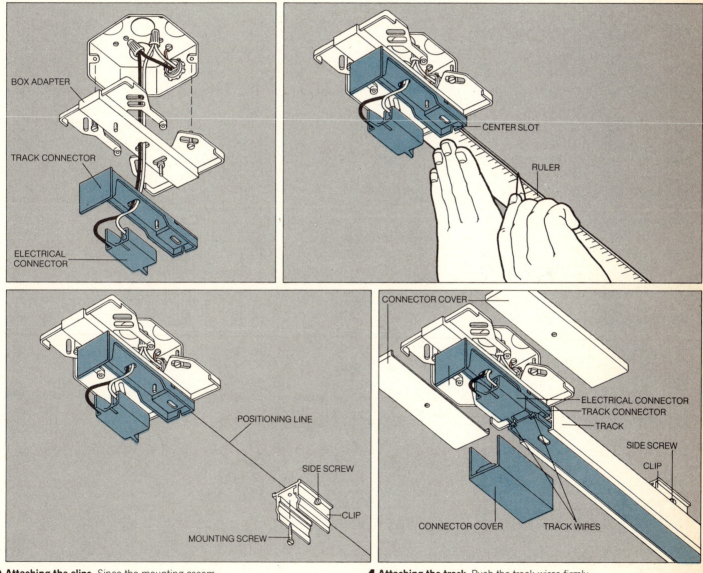

BOX ADAPTER

TRACK CONNECTOR

ELECTRICAL CONNECTOR

CENTER SLOT

RULER

POSITIONING LINE

SIDE SCREW

CLIP

MOUNTING SCREW

CONNECTOR COVER

ELECTRICAL CONNECTOR

TRACK CONNECTOR

TRACK

SIDE SCREW

CLIP

CONNECTOR COVER

TRACK WIRES

3 Attaching the clips. Since the mounting assembly holds the end of the track slightly away from the ceiling or wall, special clips are needed to keep the rest of the track level. To install the clips, make marks at even intervals along the positioning line drawn in Step 2. Drill pilot holes at those points and screw the clips to the wall or ceiling. Partially insert the side screws into the clips.

4 Attaching the track. Push the track wires firmly into the electrical connector and slide the track into the track connector. Push the track up and into the clips. Tighten the side screws in the clips and snap on the connector cover.

Doubling Up on New Receptacles and Switches

Adding new appliances and lamps in an old home usually means adding new receptacles too. Once you have made the decision to run new cable and install outlet boxes for these receptacles, you should seek the best return for your labor, either in the maximum number of receptacles or in versatile receptacles for special purposes.

There are a number of ways of wiring outlet boxes for four receptacles. You can mount two duplex, or double, receptacles in a ganged box—both receptacles are on a single 120-volt circuit. If the room has a 240-volt circuit, you can split the circuit to provide two independent 120-volt duplexes.

Whenever you install receptacles, you should consider the advantages of a circuit containing a switch that controls one or both parts of a duplex receptacle. With such a circuit, you can turn on a lamp from the entrance of a room that has no ceiling light or, with a slightly different circuit, control both a lamp and a radio by a single flick of a switch. In both circuits, the electrical connections you make will depend on whether you have a middle-of-the-run switch or a switch loop (pages 99 and 100) to work with.

Two Good Ways to Multiply Receptacles

Two duplex receptacles in one box. Gang two standard outlet boxes to make a single box (page 65), run three-wire cable to the location you have chosen for the new receptacles (page 76), attach the cable to the ganged box (page 84) and install the box in the wall (page 66). Connect the black wire of the cable to a brass terminal on one receptacle and the white wire to a silver terminal. Run a 4-inch black jumper wire from the other brass terminal to a brass terminal on the other duplex receptacle. Similarly, connect silver terminals on both duplex receptacles with a 4-inch white jumper wire. By ganging more boxes you can add more receptacles in the same manner.

To ground the installation, attach two 4-inch green or bare jumper wires to the green terminals of the duplexes, attach a third jumper wire to the box with a machine screw, and connect these jumpers to the bare cable wire with a wire cap.

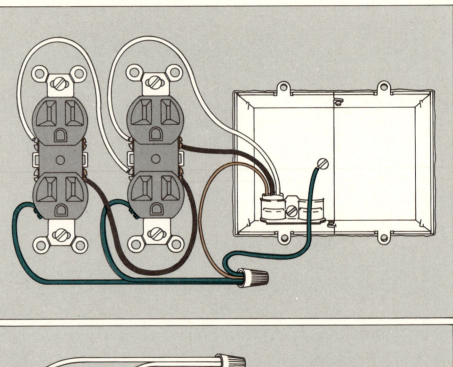

120-volt duplexes in a 240-volt circuit. Gang two standard outlet boxes into one (page 65) and connect the 240-volt, four-wire cable to the ganged box. Run the black wire of the cable to a brass terminal on one duplex and the red wire to a brass terminal on the other. Attach 4-inch white jumper wires to the silver terminals on the two duplexes and connect these jumpers to the white cable wire with a wire cap. Then ground the installation by the method described above.

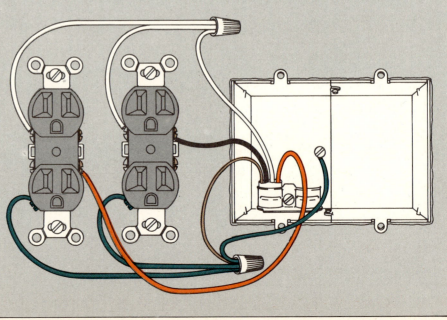

A Switch to Control a Duplex Receptacle

A switch loop. This scheme is used where it is simpler to bring the incoming power cable to the receptacle than to the switch. At the receptacle box, use a wire cap to connect the incoming black wire to the black wire of an outgoing cable that runs to the switch box. Attach the incoming white wire to a silver receptacle terminal, the outgoing white wire to a brass terminal. The outgoing white wire will now function as a hot wire. Identify it as hot: code it with a piece of black electrician's tape or a dab of black paint on the insulation next to the terminal. Attach a 4-inch green or bare jumper wire to the green receptacle terminal, and attach a second jumper to the back of the box with a machine screw; connect the jumpers and the two bare cable wires with a wire cap.

At the switch, attach the black wire to one of the brass switch terminals, the white wire to the other. Code the white wire black. Finally, pigtail two short bare or green jumper wires to the bare cable wire. Attach one jumper to the back of the box with a machine screw; run the other jumper to the green grounding terminal on the switch.

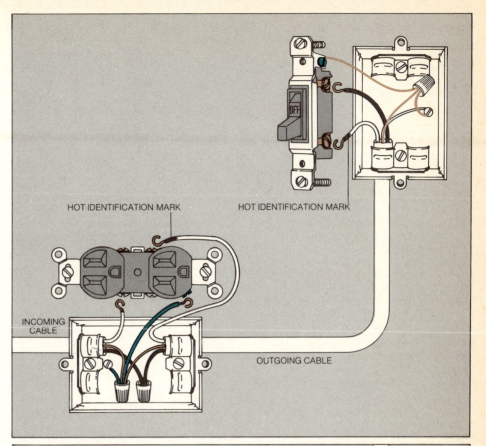

HOT IDENTIFICATION MARK

HOT IDENTIFICATION MARK

INCOMING CABLE

OUTGOING CABLE

A middle-of-the-run switch. Where it is more convenient to bring the incoming power cable to the switch box than to the receptacle box, do so and add a cable extending the circuit to the receptacle box. At the switch box, connect the white wires of the two cables with a wire cap and attach the black wires to the brass terminals of the switch. Attach a 4-inch green or bare jumper wire to the box with a machine screw. Attach a second jumper to the grounding terminal on the switch. Connect these jumpers to the bare cable wires with a wire cap.

At the receptacle box, connect the black wire to a brass receptacle terminal and the white wire to a silver terminal. Attach a 4-inch jumper wire to the back of the box with a machine screw and another to the green receptacle terminal. Then connect the two jumper wires to the bare cable wire with a wire cap. In this type of installation all white wires remain neutral.

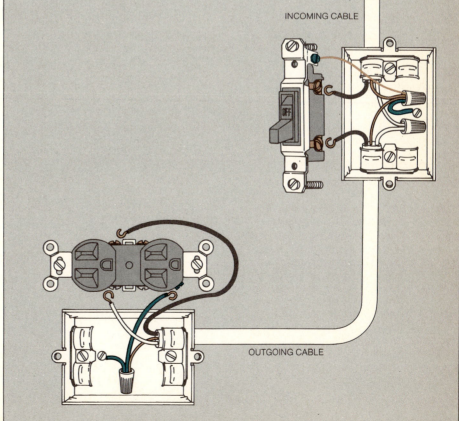

INCOMING CABLE

OUTGOING CABLE

A Switch to Control Half a Duplex Receptacle

Dividing the receptacle. To divide a duplex receptacle so that one is controlled by a switch and the other remains hot at all times, the receptacle must be modified. Break the connection linking its two brass terminals by snapping off the tab in the center *(right)*. Then use one of the methods described below, which differ with the kind of installation being made. (Here, the switch controls the right half of the receptacle pair, and the left half is independent.)

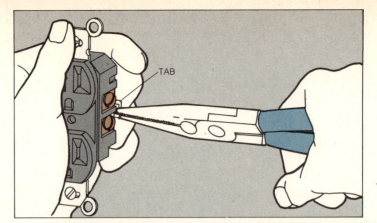

TAB

Switch loop. Bring the incoming three-wire cable to the receptacle box and run another length of three-wire cable to the switch box. At the receptacle, connect the incoming white wire to a silver terminal of the modified receptacle, and the outgoing white wire to a brass terminal. The outgoing white wire will function as a hot wire; code it black. Attach a 3-inch black wire to the other brass terminal and connect it to the black cable wires with a wire cap. Attach a 4-inch jumper wire to the green receptacle terminal, another jumper to the back of the box with a machine screw. Join the jumpers and the bare cable wires with a wire cap.

At the switch box, attach the black cable wire to one switch terminal, the white wire to the other. Identify the white wire as a hot wire with tape or black paint. Pigtail the bare wire to two short jumper wires. Secure one to the back of the box with a machine screw, the other to the green terminal of the switch.

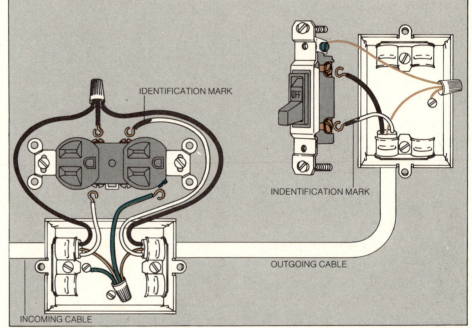

IDENTIFICATION MARK

INDENTIFICATION MARK

OUTGOING CABLE

INCOMING CABLE

A middle-of-the-run switch. Bring the incoming three-wire cable to the switch box. The connection from the switch box to the receptacle box must be four-wire cable, with a red and a black hot wire. At the switch box, join the white cable wires with a wire cap. Attach a 3-inch black jumper to one switch terminal; connect it to the incoming black and outgoing red wires with a wire cap. Attach the outgoing black wire to the other switch terminal. Attach a 4-inch jumper wire to the back of the box with a machine screw. Attach a second jumper to the switch's grounding terminal. Join these jumpers to the bare cable wires with a wire cap.

Connect the red and black wires of the cable to the brass terminals of the modified receptacle. Connect the white wire to a silver terminal. Then attach one 4-inch green or bare jumper to the green receptacle terminal screw, a second one to the back of the box with a machine screw. Connect the jumpers and the bare cable wire with a wire cap.

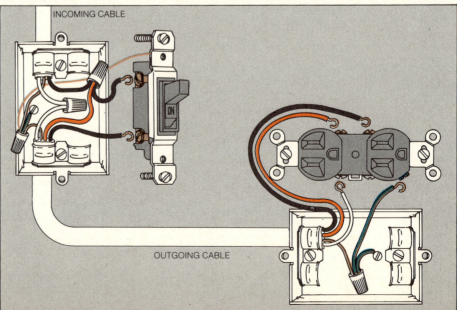

INCOMING CABLE

OUTGOING CABLE

Surface Wiring: A Shortcut to New Circuits

There is a way to install new receptacles, switches or fixture boxes exactly where you want them without running cable through walls: use raceway. It is a system of channels and outlet boxes that extends a circuit from an existing box along the surface of a wall or ceiling. It is like a permanently installed extension cord but neater—the channels can often be arranged to look like part of a baseboard or window molding—and safer, because the rigid installation prevents mechanical damage to wires and, in the metal type, provides a ground connection.

Several kinds of raceway are available. The type shown on the following two pages is the least likely to damage the insulation of the wires as you thread them through. It has channels that resemble flattened tubes, which snap into clips fastened to the wall. After inserting a bushing at each end of the channel, Type TW wires *(page 24)* are threaded through the channel as they are through conduit.

Another type of raceway has a two-piece channel—the back is fastened to the wall, the wires are laid into place and the front is snapped on. Systems using prewired plastic strips are also made, but most of them lack any provision for grounding and need special fittings. Metal raceway systems are installed using elbow and T connectors to route channels around corners; they also can use surface-mounted boxes, which accept standard receptacles, fixtures and switch-es that are used with in-the-wall wiring.

The size of the raceway channel you need depends on the number of wires required. The smallest size, designed for three wires, is usually adequate.

Whatever hardware you use, plan your installation carefully *(below)*. First locate the position of the outlets you want to install, then find the most convenient existing outlet to connect them to. To add raceway receptacles you must tap an existing outlet that has an incoming power line. But you can also use raceway for jobs that may not require direct connection to an incoming power line, such as moving an existing fixture or switch, or converting an existing switch into a pair of three-way switches.

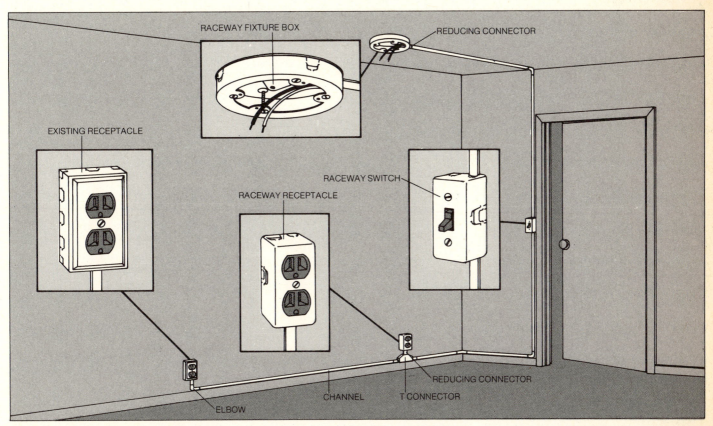

RACEWAY FIXTURE BOX

REDUCING CONNECTOR

EXISTING RECEPTACLE

RACEWAY SWITCH

RACEWAY RECEPTACLE

REDUCING CONNECTOR

CHANNEL

T CONNECTOR

ELBOW

A plan for raceway. Sketch the layout you intend to install and transfer your sketch to the walls and ceiling so that you will be able to locate parts, measure distances and estimate materials that will be needed. Follow this plan during the actual installation, since it must proceed in sequence using the hard-ware shown on the following pages. First turn off the power. Put in the extension to tap an existing outlet—a receptacle, in the layout shown above. Then fasten mounting plates for raceway receptacles, switches and fixtures. Next attach mounting clips for the channel, and the base plates of elbows, Ts and reduc-ing connectors that join parts of different sizes. When all of the bases and mounting plates are in place, measure, cut and install the channels. Only then should you thread wires through the channel, make all wiring connections and attach the various devices to their mounting plates.

The Step-by-Step Plan
for a Raceway System

Extending the existing box. A two-part adapter, rectangular or round to fit wall or ceiling outlets, extends a regular box outward so raceway can be connected to it. To install one at a receptacle, turn off the current and remove the receptacle from the box but do not disconnect the wiring. Slip the adapter's tongued plate over the receptacle. Prepare the adapter extension frame to accommodate the channel, using pliers to remove a twist-out slot where the channel will enter. Then slip the extension frame over the receptacle and clip the channel to the desired tongue; screw the frame and the tongued plate to the box.

Wires are installed after the channel is in place. Connect wires, depending on the purpose of the installation, by the same method as shown for in-the-wall installations (*pages 54-55*).

The same extension adapter is used for tapping switch boxes as for receptacles. For connecting raceway to existing light-fixture boxes, an adapter like the raceway fixture box (*opposite*) is used.

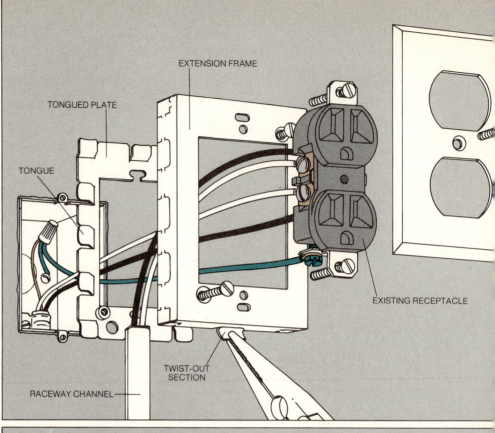

EXTENSION FRAME

TONGUED PLATE

TONGUE

EXISTING RECEPTACLE

TWIST-OUT SECTION

RACEWAY CHANNEL

Raceway receptacle and switch. The same kind of base is used for a raceway receptacle (*right*) or switch: a metal mounting plate with twist-off tongues along the edge. Remove all the tongues except the one where the raceway channel joins this unit. Attach the plate to the wall with screws, using anchors in plaster or wallboard.

Measure and cut the raceway channels. Wires are attached much the same way as for a receptacle or switch after the channel is in place, clipped to the tongue left on the mounting plate. The difference is in grounding. Attach one end of a 6-inch jumper to the green terminal and the other end to one of the screws securing the mounting plate. Screw the receptacle or switch to the plate. Remove a twist-out from the cover so it fits over the raceway channel and screw on the cover.

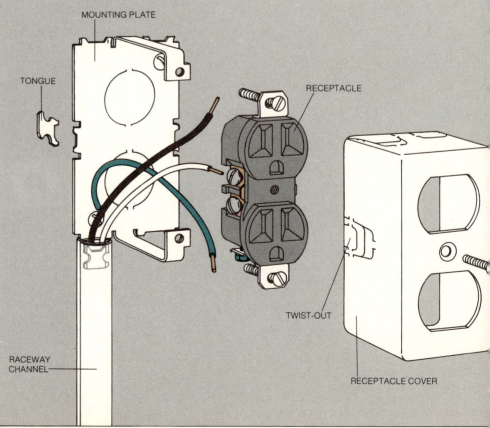

MOUNTING PLATE

TONGUE

RECEPTACLE

RACEWAY CHANNEL

TWIST-OUT

RECEPTACLE COVER

Raceway fixtures. A raceway light-fixture box is installed in the same way as a switch or receptacle *(previous page)* with one exception: A reducing connector is needed to join it to the small-sized raceway channel that takes three wires. Attach the reducing-connector base so the larger end of it overlaps the tongue on the fixture mounting plate. Push the connector cover onto the base. To attach the raceway channel to the connector, push it onto the tongue.

Fitting in raceway channel. Using a fine-tooth hacksaw (40 teeth per inch), cut each section of raceway channel to exactly the length between the base of the tongue on one unit to the base of the tongue on the next unit, such as the flat elbow connector and the tongued plate shown directly below. If you need a section longer than any piece you have, join two with an extension connector *(bottom)*. To install a channel section, push one end onto a tongue; then loosen, or if necessary, remove screws holding the unit opposite so that you can join its tongue to the other end of the raceway channel.

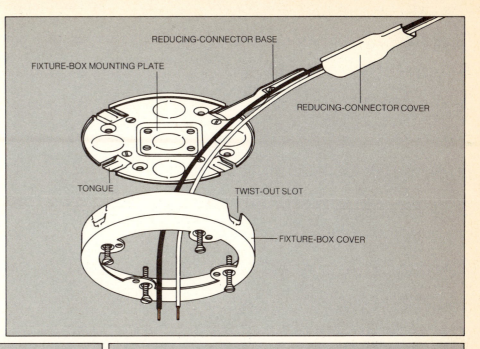

REDUCING-CONNECTOR BASE

FIXTURE-BOX MOUNTING PLATE

REDUCING-CONNECTOR COVER

TONGUE

TWIST-OUT SLOT

FIXTURE-BOX COVER

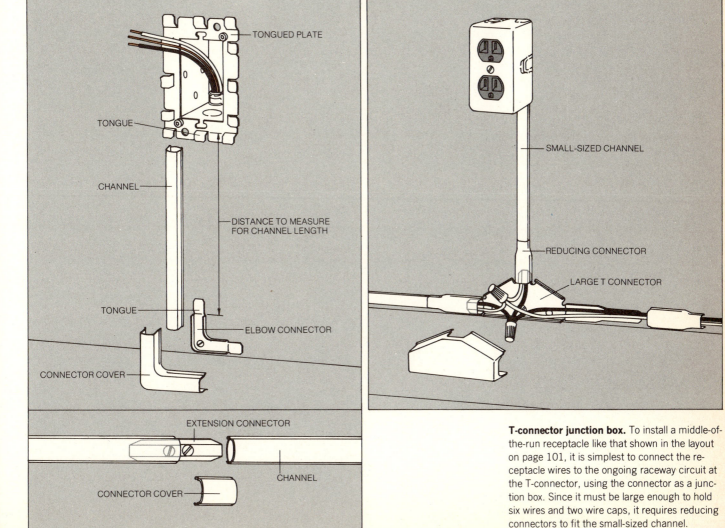

TONGUED PLATE

TONGUE

CHANNEL

DISTANCE TO MEASURE FOR CHANNEL LENGTH

TONGUE

ELBOW CONNECTOR

CONNECTOR COVER

EXTENSION CONNECTOR

CHANNEL

CONNECTOR COVER

SMALL-SIZED CHANNEL

REDUCING CONNECTOR

LARGE T CONNECTOR

T-connector junction box. To install a middle-of-the-run receptacle like that shown in the layout on page 101, it is simplest to connect the receptacle wires to the ongoing raceway circuit at the T-connector, using the connector as a junction box. Since it must be large enough to hold six wires and two wire caps, it requires reducing connectors to fit the small-sized channel.

The Final Touch: Patching Holes

Installing new wiring usually leaves holes in walls and ceilings. They may be as small as a ⅛-inch test hole or as big as the gap left by a box opening cut in the wrong place. In all but exceptional cases, a simple repair restores the surface to its original condition.

The method depends upon the size of the hole and the composition of the surface. Use spackling compound to fill test holes in plaster or plasterboard; in wood, use plastic wood that matches the wall. For slightly bigger openings such as the open seams around a newly installed box, use plastic wood or patching plaster but stuff the openings with steel wool to support the patches *(top right)*.

Holes up to 3 or 4 inches, such as the access holes used for running cable or for an opening made by mistake, are patched with new sections of plasterboard, plaster or wood. If the hole lies directly over a stud or joist, you can use these structural beams as a support for the patch *(bottom right)*. To fill an opening between two studs or joists, you will have to make a backing section of plasterboard, wood or—for a plaster wall —metal screening *(opposite)*.

Even large holes in plasterboard are easy to repair. Expand the hole by removing an entire section around it as far as the nearest studs or joists, then replace the section with new plasterboard *(pages 73-74, Steps 1-5 and 7)*. Big holes in plaster or wood paneling, however, may require repair by a professional—plaster patches tend to crack and crumble, and wood ones must be applied skillfully in order to avoid gaps.

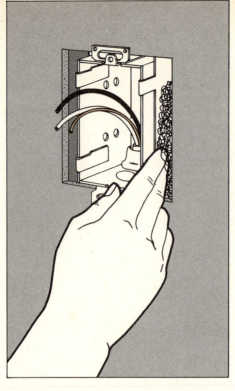

Filling the Gaps around an Outlet Box

Making a steel-wool backing. Pack handfuls of steel wool into the gaps at the edges of the box. In a wood surface, fill the gaps flush to the surface with matching plastic wood and sand the patch smooth. In plaster or plasterboard, mix a batch of thick patching plaster and moisten the outer edges of the gaps. Beginning at these edges and working toward the box, apply plaster to a point just below the surface with a putty knife. Score the wet plaster with the tip of the putty knife. Let the plaster dry, then fill the gaps with plasterboard joint cement, and smooth the cement about 2 inches outside the gaps with a broad-bladed tool called a taping knife. Allow several days for the patch to dry, then sand.

Repairing Holes at Studs and Joists

Making a plasterboard patch. In a plasterboard wall or ceiling, measure the exact dimensions of the hole and cut a new section of plasterboard ⅛ inch smaller. Insert the patch with the lighter side facing outward and attach it to the stud or joist with two plasterboard nails. Drive each nail flush with the surface, then gently hammer the nail once more to drive the head a fraction of an inch below the surface so that the rounded face of the hammer makes a small depression, or dimple, in the plasterboard. Fill the cracks around the edges of the patch and cover the dimpled nails with plasterboard joint cement.

In a plaster-and-lath surface, apply enough layers of patching plaster to fill the hole almost completely, let the plaster dry, then fill the hole flush to the surrounding surface with joint cement. When the surface is dry, sand it smooth.

In a wood surface, cut a matching piece of wood to fit the hole and nail it to the stud or joist with finishing nails. Countersink the nails. Fill the countersunk holes and the cracks around the patch with plastic wood, and sand smooth.

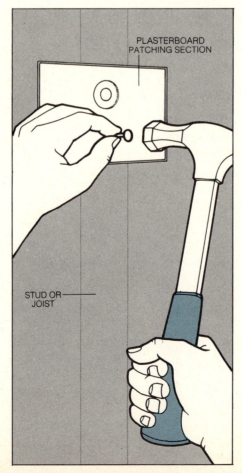

PLASTERBOARD PATCHING SECTION

STUD OR JOIST

Repairing Holes between Studs and Joists — in Plasterboard and Wood

1 Preparing the patch. For a plasterboard wall or ceiling, cut a plasterboard patching section ⅛ inch smaller than the opening (*page 74, Step 5*), and a backing section about 1 inch narrower and 2 inches longer. Set the patching section across the backing section, with the dark side of the patching section facing in, and fasten them with joint cement. When the cement dries, drill a ¼-inch hole through the center of the sections. Tie the ends of a 6-inch string to a nail, push the string loop through the hole in the backing section and pull the loop out the other side.

For wood surfaces, follow the same procedure with sections of wood, but cut the patching section to a precise fit with the opening and use white glue as the adhesive for the patch.

2 Installing the patch. Apply plasterboard joint cement or white glue to the parts of the backing section that project beyond the edges of the patching section. Holding your finger through the loop of the string, insert the bonded sections horizontally through the opening and into the space behind. Then turn the sections to the vertical and pull them toward you until the patching section fills the opening.

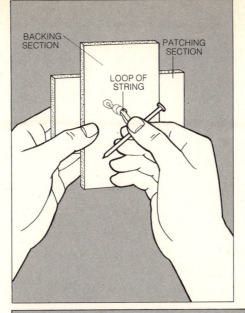

BACKING SECTION
PATCHING SECTION
LOOP OF STRING

3 Securing the patch. Slip a pencil through the loop of the string and turn it until the string is taut and the pencil pulled tight against the patch. Turn the pencil until its ends extend beyond the patch, locking it in place. When the adhesive dries, remove the pencil and push the string through the hole into the wall.

In plasterboard, fill the cracks at the edges of the patch and the hole at the center with joint cement and let it dry; apply as many coats as you need to create an even, smooth surface and sand each coat smooth. In wood, fill the cracks and hole with matching plastic wood, let the wood dry and sand it smooth.

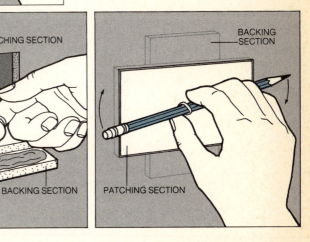

ADHESIVE
PATCHING SECTION
BACKING SECTION
ADHESIVE

BACKING SECTION
PATCHING SECTION

Repairing Holes between Studs and Joists — in Plaster

Applying plaster to a metal screen. Cut a piece of stiff metal window screening slightly larger than the hole and thread the ends of a length of string through the middle. Moisten the inside edges of the hole and apply patching plaster to these edges; reach inside the hole and spread patching plaster around the back surface of the wall. Holding the free ends of the string, insert the screen into the hole, then use the string to pull the screen flat against the back of the wall, embedding it in the wet plaster. Tie the string around a pencil and turn the pencil to tighten the string and hold the screen firmly in place. Fill the hole with plaster to about ¼ inch below the surrounding surface and score the surface of the wet plaster. When the plaster sets, cut the string off at the surface; finish with a coat of plasterboard joint cement and a thorough sanding.

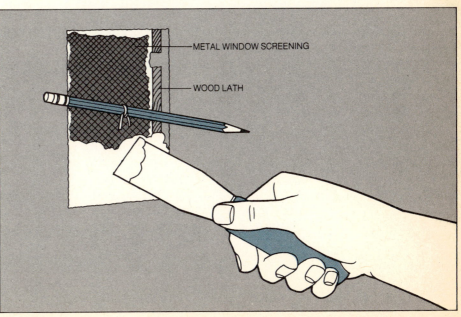

METAL WINDOW SCREENING
WOOD LATH

Low Voltage for Door Chimes

Doorbell and chime wiring is one of the few electrical systems in a house that carry less than 120 volts. No chimes require more than 20 volts, and such a low voltage means that installing chimes is, in most respects, simpler and quicker than other wiring jobs.

A transformer—except for its size, similar to the one on a utility pole *(page 12)*—is attached to a regular outlet box to reduce 120-volt house current to the level required. Small-diameter wires (18 or 20 gauge) called bell wires connect the transformer to the chimes and push button. Electrical codes prescribe no special wiring rules for doorbells except that the transformer must be connected to the house current at a regular box. All other connections can be made without boxes. There are no grounding requirements and no distinction between hot and neutral wires, so there is no universal color code. If you are installing new wiring, however, choose three different colors for the wires to help you make correct connections.

The drawings and instructions on these pages explain how to install an entire door-chime system, including the wiring, but the same procedures are used for connecting new components to existing wiring. The transformer is commonly installed in the basement and chimes are often mounted on a hallway wall, where they are out of the way and can be heard through the house.

Extend a circuit for the transformer, then install the bell wires. Though these wires can be strung in the open, they should be placed inside walls, attics or basements wherever possible. First drill holes behind the push buttons and the chimes. Then use the techniques illustrated on pages 76-83 to fish two wires by a convenient route (from the basement in this example) through each of the push-button wiring holes. Run the other end of one wire from each push button to the chimes. Run the other push-button wires to the transformer, and bring a third wire from the transformer to the chimes. Connect the bell wires, then turn off the power before connecting the transformer to the house circuit.

Installing a push button. Bore a ⅝-inch hole at about doorknob height into the hollow space of the outer wall of the house, locating the hole 4½ inches from the edge of the door to miss the double studs supporting the doorway. Position the push button at the wiring hole. Mark and drill pilot holes for the screws that will secure the push button to the wall. After wires are installed and fished through—one from the chimes and one from the transformer—attach one wire to each screw terminal and mount the button.

Mounting chimes. Use the chime mounting plate to locate holes for screws and wires. Bore the holes, and after installing the wires and fishing three through the hole—one from each push button and one from the transformer—screw the chimes to the wall. Attach the transformer wire to the terminal marked TR, the front-door push-button wire to the terminal labeled FT and the back-door push-button wire to the remaining terminal. Then position the wires so that they will not interfere with the plungers.

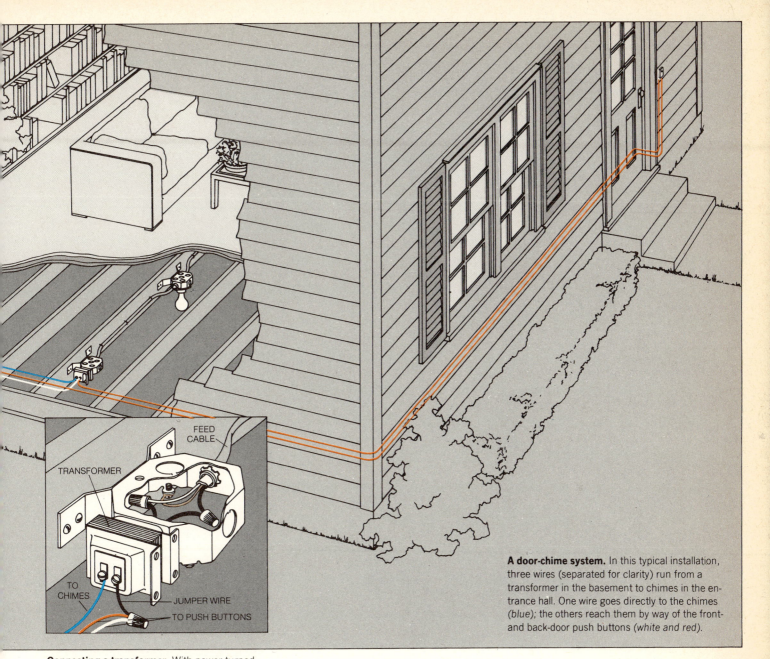

FEED CABLE

TRANSFORMER

TO CHIMES

JUMPER WIRE

TO PUSH BUTTONS

A door-chime system. In this typical installation, three wires (separated for clarity) run from a transformer in the basement to chimes in the entrance hall. One wire goes directly to the chimes (*blue*); the others reach them by way of the front- and back-door push buttons (*white and red*).

Connecting a transformer. With power turned off, extend a circuit (*page 76*) to a junction box mounted on a basement floor joist. Remove a knockout from one side of the box, pass the transformer wires through the hole, then lock the transformer to the outside of the box by fastening a star nut to the transformer's threaded shaft. Connect one transformer wire to the black house cable and the other to the white cable.

Connect the wire from the chimes to either of the transformer terminals. Connect a jumper to the other transformer terminal and, using a wire cap, connect the jumper to the wires that lead to the push buttons. The jumper wire may be as long as necessary to reach a convenient point for the push-button wires to branch off.

Bringing Electricity Outdoors

Darkness interrupts a patio dinner before dessert is served. A 75-foot extension cord drags awkwardly behind a hedge trimmer. Winter ice along the edge of a roof lifts shingles and loosens gutters. There is a remedy for a home with such problems and inconveniences: extend a circuit from inside the house to bring electricity outdoors. The power can be used for lights to extend the hours of backyard festivities. It can supply outdoor receptacles that make long extension cords unnecessary, or it can energize heating cables to keep water from freezing and damaging a roof.

The first step to be taken before undertaking any outdoor wiring should be a visit to your city or county building department to obtain a permit, if one is required, and to find out if there are any local requirements for exterior wiring. The procedures that are discussed on the following pages are in accord with the electrical codes adhered to generally in the United States and Canada.

Indoor and outdoor circuitry are electrically identical, so the basic indoor wiring techniques presented in the previous three chapters are the same ones used to extend wiring outdoors. Differences between the two kinds of wiring lie mainly in the specialized equipment and materials necessary to install wires outside. Interior and exterior wires, for example, are made of the same metals and share the same color-code. But because outdoor conductors usually run underground, they must be sheathed in tough plastic or enclosed in conduit if they are to endure years of burial in damp earth. Indoor receptacles are used outdoors, but they must be installed in weatherproof boxes that keep out rain and snow. Circuits inside the house and out are equipped with circuit breakers or fuses, but the National Electrical Code requires that all outdoor lines containing receptacles have the added protection of a special kind of circuit breaker, called a ground-fault interrupter *(page 116)*, that makes a serious shock from an outdoor appliance virtually impossible.

Installing an outdoor circuit calls for drilling a hole through a wall of the house for wires to pass outside; then the opening must be sealed against moisture. It will be necessary to dig a trench, whose depth will vary, depending on whether cable is to be buried directly in the ground or individual wires protectively encased in conduit. In every outdoor installation, short stretches of conduit are necessary wherever cable or wires are exposed aboveground. Holes for lampposts must be sunk and the posts anchored to keep them upright. Heating cables need to be fastened to the roof, and freestanding receptacles often require a cinder block to help support them. These chores are worth the effort, however, making a yard safer and more enjoyable and perhaps increasing the value of the property as well.

Planning the Circuits

Most outdoor wiring jobs have two phases. One is the installation of indoor wiring to carry electricity as far as the point where it leaves the house; this part of the job is described on pages 76-83. The second phase, the installation of the outdoor wiring, differs in several important respects. You must decide where the wires will emerge from the house, determine how much conduit and what kind of wire to use outdoors, and plan the route of a trench for the wiring.

A sketch of your house and yard will help you lay out wires, receptacles and fixtures, and it can later serve as the basis for a list of materials and tools. In making the sketch, first decide upon the placement of light switches and fixtures for decoration, for safety or to discourage prowlers. Some installations have simple single-pole switches that control fixtures from one spot. For extra convenience, consider an arrangement of three-way switches to control lights from inside and outside the house, or a photoelectric switch, which automatically turns lights on at dusk and off at sunrise.

Next indicate the best route to the indoor connection and select the location for a ground-fault interrupter (page 116) to prevent shock. Choose a circuit that has sufficient unused capacity for the number of outlets you plan to install (page 20). Often a porch light or an existing outdoor receptacle can serve as the starting point for new wiring (page 118). Cellar and attic lighting circuits are also good candidates: many of them are underutilized, and are easy to work on because the wiring is usually exposed indoors and is simple to bring outdoors through the foundation, a floor joist, a wall or an eave (page 119).

Before you bring wire outside your house, you must first decide what kind of wiring to use. This decision depends in part upon your local electrical code. Most codes require that outdoor wiring must run within conduit from the point it leaves a house to the point it disappears underground. Similarly, it must be protected by conduit wherever it rises above ground to an outlet box.

Beyond this area of agreement, local codes differ widely on the extent to which conduit must be used and the kinds of conduit that are acceptable. In many locations, most underground wiring can be UF cable, a plastic-sheathed cable designed to be buried directly in the earth. In other areas, conduit must protect all underground wiring. In such situations, UF cable is rarely used; instead, individual, type TW wires, having moisture-resistant plastic insulation, are fished through the conduit.

Three types of conduit are commonly used outdoors: rigid plastic; thin-wall metal (or EMT—electrical metallic tubing); and rigid, or heavy-wall, metal. The choice among them and UF cable depends on an individual evaluation of cost and effort. All are installed with similar techniques. UF cable is the least expensive. It, as well as plastic and thin-wall conduit, is relatively easy to work with, but all require that you dig a trench at least a foot deep. Rigid conduit, on the other hand, is more costly and takes some effort to bend, but it can be buried in a relatively shallow, 6-inch trench; to save digging, many homeowners prefer it, and it is the type that is shown on the pages that follow.

The trench for cable or conduit must avoid, as far as possible, such underground obstacles as electric and telephone cables, water lines, sewer connections and sprinkler systems. Do not run underground wiring of any kind through the corrosive chemicals of a septic-tank drainage field or in areas that are frequently covered by puddles of standing water. Probe for rocks and other obstructions along the planned route of the trench before you start digging.

With your plan of attack drawn up, make a list of all the fixtures, wires, cable, conduit and tools you will need. Assemble all tools and materials before beginning work to avoid interruptions for trips to the hardware store.

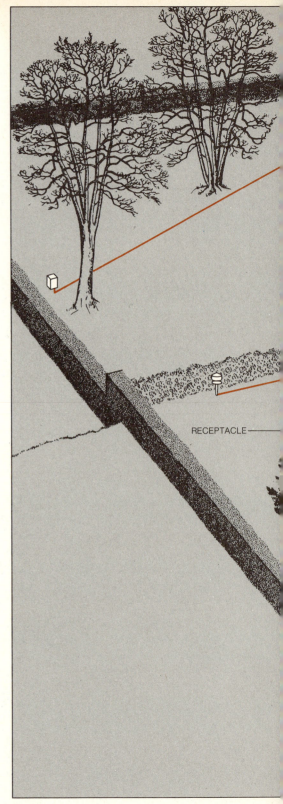

RECEPTACLE

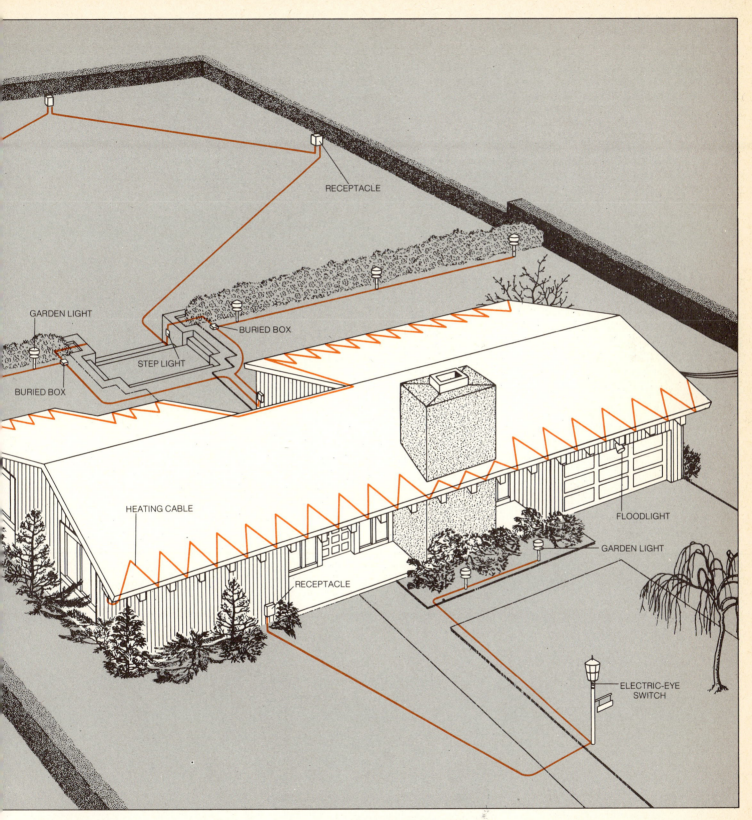

RECEPTACLE

GARDEN LIGHT

BURIED BOX

STEP LIGHT

BURIED BOX

HEATING CABLE

FLOODLIGHT

GARDEN LIGHT

RECEPTACLE

ELECTRIC-EYE SWITCH

An outdoor wiring scheme. At the front of the house, a typical outdoor circuit begins with a wall-mounted receptacle, then goes underground to a post lamp controlled by an electric-eye switch. The circuit continues to a pair of garden lights. Wiring from a separate exit feeds a floodlight above the garage door. Behind the house, a circuit branches as it comes through the wall. The right side forks at a buried box to serve garden lights, a step light and three outlying receptacles. (Boxes may be buried only if their location is effectively identified and they are accessible for excavation.) The left side provides power for a second step light and two more garden lights. Heating cables from the roof plug into receptacles under the eaves.

Weatherproof Fixtures

Electrically, the parts for outdoor wiring —switches, fixtures, boxes and other fittings—are identical to those used indoors. Mechanically, they may not be. Like the components on these pages—a right-angle LB fitting that routes wires out of the house, an outdoor box and receptacle cover plate, an outdoor switch and a light fixture—outdoor parts are either built to be impervious to rain and snow, or designed for assembly in a weatherproof container.

The outlet boxes for these components are made of heavy metal castings. Instead of knockouts for incoming and outgoing wires, the boxes have threaded holes into which metal plugs or the ends of conduit are screwed. The cover plates are secured by screws that fit snugly into countersunk holes, and gaskets keep moisture from seeping underneath the plates. Even these protective devices are inadequate if the outlet box is installed where it might be submerged by a heavy rain or thawing ice. If such locations are unavoidable, the homeowner must pay the extra cost of hardware fitted with watertight seams.

The LB fitting. This connector (LB indicates an L-shaped fitting with a back conduit opening) is installed where wires pass through an outside wall of a house, to turn them toward the ground. The hole at each end is threaded for conduit, and an access plate, secured by two screws and equipped with a gasket for weatherproofing, makes it easy to fish conductors through. The LB fitting is too small to hold splices; it can be used only to turn continuous wires around a corner.

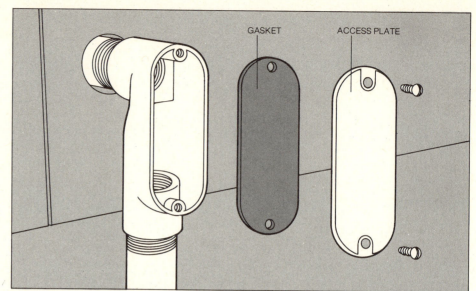

GASKET ACCESS PLATE

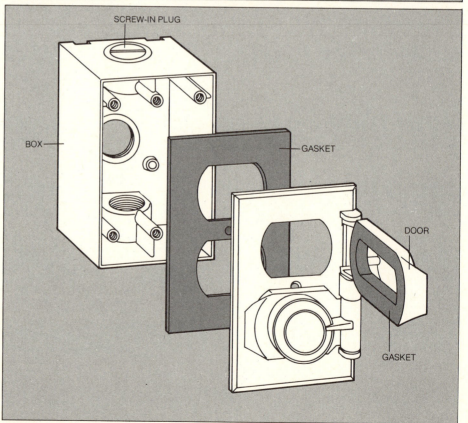

SCREW-IN PLUG

BOX

GASKET

DOOR

GASKET

Weatherproofing a receptacle. A special cover plate makes it possible to install an ordinary indoor receptacle in an outdoor box. Two gasketed doors seal the receptacle sections when they are not in use. A larger gasket seals the gap between the cover plate and the rim of the box.

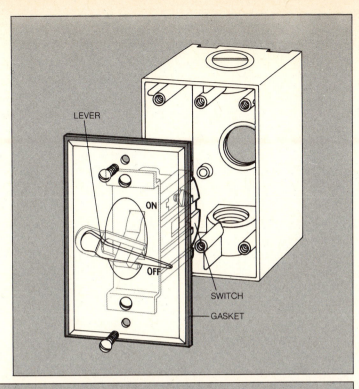

An outdoor switch. This device comes preassembled with a weatherproof cover plate for an outdoor box. An external lever extends through the plate to trip a toggle switch inside the box.

LEVER

ON

OFF

SWITCH

GASKET

A weatherproof light fixture. To keep water out of an outdoor light socket, a gasket seals the gap between the side of the fixture and the base of the bulb. Another type also has a metal shroud to protect the bulb. Each fixture is locked to the cover plate with a star nut. Other types of cover plates—including some for larger or different-shaped boxes—are also available to mount two or three sockets on one box. All such fixtures require weatherproof bulbs made of glass that will not shatter if cooled suddenly by rain or snow.

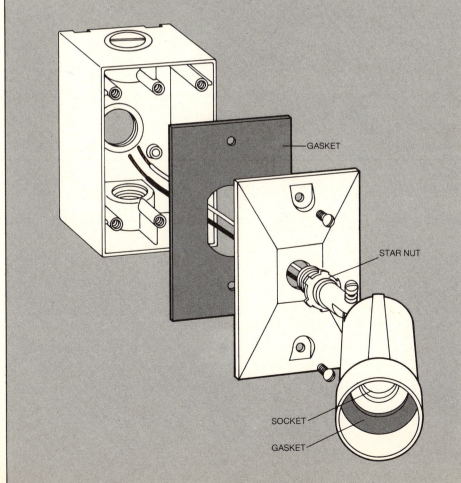

GASKET

STAR NUT

SOCKET

GASKET

Working with Heavy-Wall Conduit

Rigid metal conduit is made of either aluminum or galvanized steel. Use aluminum if your local code permits you to: it protects wiring as well as steel does, but costs no more. It is easier to bend and cut, and about a third the weight of steel. You must take one precaution, however, if you plan to set aluminum conduit in concrete: the conduit must first be protected against corrosion by coating it with bituminous paint.

Conduit with an inside diameter of ½ inch is large enough for most home wiring. In a 15-ampere outdoor circuit, it can carry one 14-gauge, 3-wire UF cable or nine 14-gauge TW wires. Aluminum and steel ½-inch conduit is normally available in 10-foot lengths with threaded ends, and each section comes with a threaded coupling to join it to another. Steel conduit is also available in the form of nipples—short lengths, from ¾ inch to 36 inches long, that are threaded on both ends but are not fitted with couplings. The steel nipples are used with aluminum as well as steel conduit. A variant of the nipple is an elbow, also threaded on both ends, but bent to a 90° curve and available in both aluminum and steel.

Most work with conduit is done using everyday tools: a hacksaw for cutting, a round file to deburr rough edges, and a C clamp for locking conduit to a sawhorse while you cut. But you also need a few specialized tools. A conduit bender is needed to shape straight lengths of conduit. You should also have a pair of 10-inch pipe wrenches to screw conduit components tightly together, and a 12-inch-long adjustable wrench (or a 1⅛-inch open-end wrench) to fit the nuts on the special connectors that are needed for the threadless ends of trimmed conduit *(opposite)*—only a wrench that large will serve.

Assemble conduit alongside the trench you have dug for it, cutting and bending sections as you need them so that you can correct small mistakes in measurements as you go. Keep conduit runs as straight as you can. If a serpentine run of conduit contains turns totaling more than 360°, interrupt the conduit with at least one box or a fitting with an access plate; otherwise, the wire will bind at turns so that you cannot fish it through. As you complete each substantial part of the assembly, lower it into the trench and connect it to the part already in place. Conduit installed along a wall must be strapped in position every 10 feet, as well as within 3 feet of each box.

Conduit components. A corner elbow, one or more conduit bodies, an offset and a few plastic bushings are all the fittings you are likely to need to install conduit. Offsets help jog conduit past small obstacles, and the screw-on plastic bushing keeps insulation from chafing against sharp conduit edges. Elbows and conduit bodies have access plates that can be removed for fishing wire. The elbow's curved access plate must be weatherproofed with caulking cord; the flat access plate of a conduit body has its own weatherproof gasket. Conduit bodies come in a variety of shapes for different purposes. The C body (C stands for continuous feed), shown in the drawing at bottom left, is used if many turns come between boxes, while a T body branches a circuit. If a corner elbow or body—or an outdoor box—is buried, its location must be permanently marked with a fixture or a short stake.

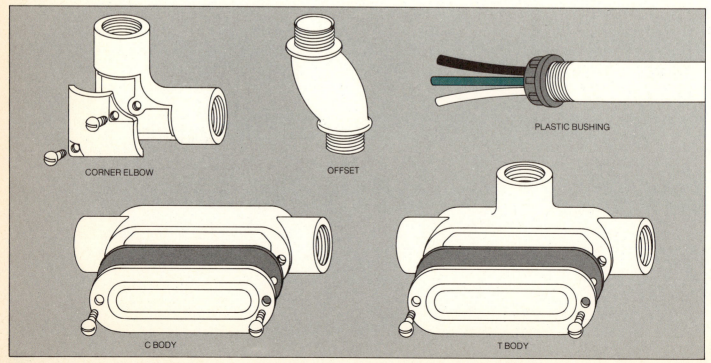

CORNER ELBOW

OFFSET

PLASTIC BUSHING

C BODY

T BODY

Conduit couplings and connectors. Couplings are used for joining two lengths of conduit, and connectors are used for fastening conduit to various pieces of hardware. Both threaded and threadless couplings join heavy-wall conduit. The threadless coupling replaces a threaded one between the unthreaded ends that are left after conduit has been cut to size. The threadless connector screws into an outlet box, a threaded coupling or a conduit component. A similar connector is also available for securing UF cable to an outlet box. Because threadless connectors are tightened by turning a nut rather than by turning an entire section of conduit, they are also used to join large, unwieldy conduit assemblies that are difficult or impossible to turn.

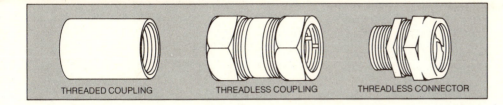

THREADED COUPLING   THREADLESS COUPLING   THREADLESS CONNECTOR

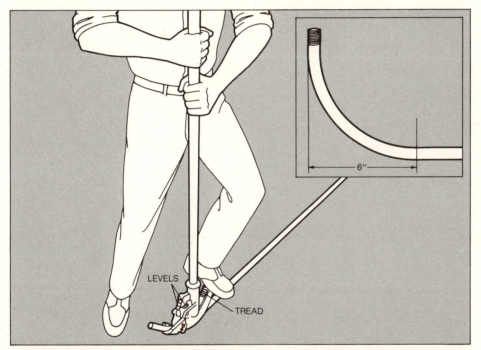

LEVELS

TREAD

6"

Making the Bends

A simple bend. Pencil a mark on the conduit where the bend is to begin, then insert the conduit into the rocker-shaped conduit bender and align the mark with the arrow on the side of the bender. Step on the rocker tread and pull on the handle. (When bending steel conduit, you will probably have to brace the opposite end against a wall.) The bender shown is fitted with two spirit levels, one to indicate that the conduit has been bent 45° and one for a 90° bend, measuring about 6 inches from beginning to end (inset). If your bender has no levels, pull the handle to a vertical position for a 45° bend; for a 90° turn, continue pulling until the handle makes a 45° angle with the ground.

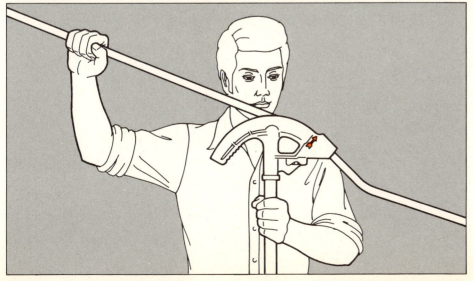

A compound bend. To jog conduit around an obstacle or fit it to the contour of a wall, make the initial bend by the method described above. Then, with the bender handle resting on the ground, position the conduit to bend in the opposite direction. Repeat the procedure to form conduit into complex shapes. Caution: You may need a helper to steady the conduit bender if you are bending galvanized steel conduit.

A Lifesaving Circuit Breaker

After mowing a damp lawn, a man stoops to unplug his electric lawn mower from an extension cord lying in the grass. As he picks up the plug, he is jolted by a severe electrical shock. Yet the mower cord was properly grounded and the circuit breaker in the service panel was in good condition. What happened?

In this incident, the man was the victim of an electrical leakage too small to trip the circuit breaker but large enough to cause harm. A 10th of an ampere—about half the current used by a 25-watt bulb—can kill by passing through the body for as little as two seconds. Usually this small current is hard to maintain in the body. But on damp earth, the man provided a path from the leaky connection to ground, and when he picked up the plug, current flowed through his body into the earth.

To prevent such accidents, the National Electrical Code requires that receptacles in new outdoor circuits and garages and bathrooms be protected with a ground-fault interrupter (GFI). Its electronic circuit compares the amperage flowing into a circuit through the black wire with the current flowing out through the white wire. If there is no leakage, the two currents are equal. But if the GFI detects a difference of .005 ampere or more—indicating a leak—it cuts off power to the circuit within $\frac{1}{40}$ of a second, so fast that serious shock is avoided.

Ground-fault protection is also required for bathrooms, garages, some basement outlets, within 6 feet of the kitchen sink and with all hot-tub equipment. Further, it is a worthwhile investment for any risky location, such as in a child's play room, a laundry room or near a wet bar.

Connecting a GFI. The GFI receptacle has wire leads instead of screw terminals but is connected like an ordinary receptacle with one crucial difference: the feed cable from the service panel must be joined to the leads marked LINE, while the outgoing cable leading to the rest of the circuit must be hooked to the leads marked LOAD. If the GFI receptacle is the only fixture in the circuit, connect the LINE leads in the ordinary way, but cover each of the LOAD leads with a wire cap and fold them into the box.

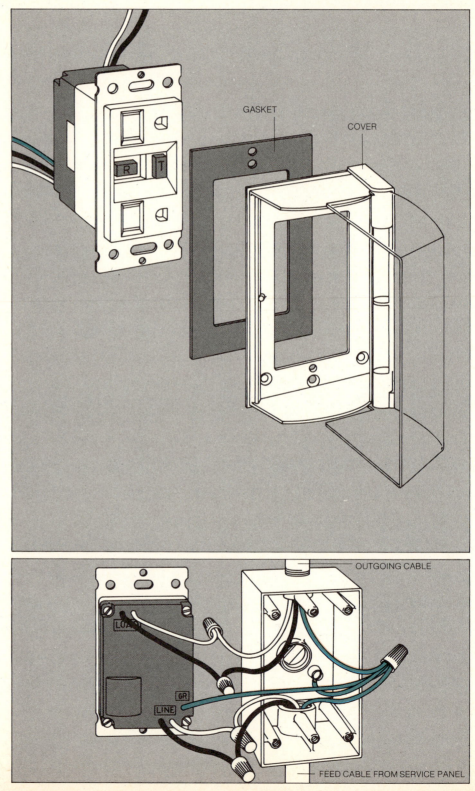

A GFI-protected receptacle. This receptacle with a built-in GFI fits an outdoor box (*bottom*) and looks much like an ordinary duplex receptacle except for two push buttons. One, labeled "R," resets the interrupter after it trips; the one marked "T" simulates a leakage to test the device.

GASKET

COVER

OUTGOING CABLE

LOAD

GR

LINE

FEED CABLE FROM SERVICE PANEL

Getting the Wires from Inside to Outside

The best way to bring electricity outdoors depends partly on the way your house is constructed, partly on whether there is an existing outlet outside and partly on whether you intend to run wiring on into the yard. The simplest outdoor installation consists of a single fixture—a receptacle or a light—mounted on an outside wall. A standard box is recessed into the wall and an indoor circuit is extended to supply power. Sinking a box into wood, asbestos or aluminum siding is essentially the same as installing it in an interior plywood wall *(page 66);* locate the box

on a single course of siding, rather than the joint between courses, and caulk the gap around it. Fitting a box into a cinder block wall *(below)* involves a few special techiques and occasionally the use of a four-edged chisel called a star drill.

To go beyond the outside wall of the house and extend wiring into the yard, you have to dig a trench and use conduit as shown on the following pages. Conduit is always needed to get wiring into and out of the trench, but not in the trench if UF cable is used.

Whether or not conduit is to be used

throughout, follow the same sequence. First, decide where and how to bring power out of the house, then drill an exit hole if needed. Below the hole, where the conduit will enter the ground, dig the trench. (If you dig the trench first, you may miss the hole with the conduit and have to dig another trench.) Assemble conduit, fittings and boxes, fish cable or wires through the conduit, fill the trench and connect the fixtures. If necessary, extend the indoor circuit by following the sequence on pages 76-83. Check the installation and restore power.

Installing a Box in Cinder Block

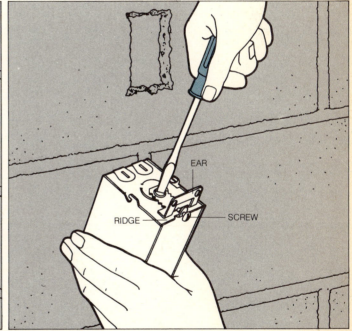

1 Making an opening. Hold the box against the center of a block and outline it with tape. Use a ⅜-inch electric drill fitted with a ½-inch masonry bit to drill several holes into the center hollow of the block, then knock out the material between the holes with a cold chisel and a ball-peen hammer. Finally, chip away the edges of the opening until it is large enough for the box.

If you cannot locate the center of a block because the seams have been stuccoed over, use a star drill to make a hole where you believe the center to be. When you have located a hollow—a spot where the star drill meets no resistance—insert a finger into the hole and feel around for the sides of the hollow. Then make an opening for a box halfway between these sides.

2 Mortaring the box in place. Insert screws loosely into the fixture-mounting tabs of the box to keep mortar out of the screw holes. Adjust the ears so that the edge of the box will extend about 1/16 inch from the wall (the cover-plate gasket will form a tight seal around this ridge), then slide the box into the cinder block. Mortar the box in place, using a putty knife. Be sure that the mortar completely fills the gap between the box and the edge of the hole so the installation will be protected from the weather. When the mortar dries, remove the screws from the mounting tabs.

Channeling Wires Underground

However you plan to extend wires into a yard, there are two basic steps that apply. You must dig a trench as described in the box on the opposite page—a shallow one for rigid conduit, a deep one for other conduit or UF cable. And you must shape conduit to carry wires from the house opening into the trench *(Step 2, below)* and on to the fixture you install *(pages 120 and 122)*. The rest of the job depends on whether you get power from an existing outlet or through a wall or eave *(opposite, bottom)*.

The simplest method is to disassemble an outside light or a wall-mounted receptacle and extend its box for conduit. However, if you tap a light, at least 6 feet of conduit will show and you may have to replace the fixture, for the extender must be capped with a weatherproof yard light. If you opt for a completely new exit hole, working from the basement or attic makes it easy to extend interior circuits. Bore a basement hole below the ground floor but not below ground level, and try to work through a joist. If you must lead wires through cinder block, go through the hollow at the center of a block in the second course of the foundation—the blocks in the top course are usually solid. In a house that has neither basement nor attic, bore the exit hole above the first floor but as close to the ground as possible.

Tapping an Existing Outlet

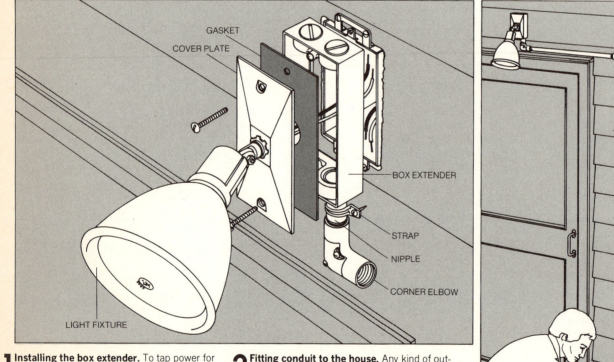

GASKET
COVER PLATE
BOX EXTENDER
STRAP
NIPPLE
CORNER ELBOW
LIGHT FIXTURE

1 Installing the box extender. To tap power for the yard from an existing exterior fixture like this light, disconnect the fixture. Assemble a corner elbow and a short nipple, then screw the combination into one of the conduit holes in an outdoor box extender. Temporarily fasten the gasket and extender to the box *(exploded drawing above)* with a yard light fixture. Then clamp the nipple to the wall with a conduit strap and caulk any gap between the box extender and the siding on the house. Later, you will take down the light and connect its leads both to the house wiring and to the wiring that you have fished through the conduit.

2 Fitting conduit to the house. Any kind of outdoor wiring must be fed from the house opening into the trench through conduit. In extending from a light, assemble conduit from the corner elbow down the wall and almost to the bottom course of siding. Extend the conduit far enough horizontally to miss the porch when it turns downward. Strap the conduit to the wall. Then, with a conduit bender, shape a conduit section to curve around the bottom of the siding and rest on the floor of the trench. Caution: Check the fit after making each bend. With a threadless connector, fasten the contoured section to the conduit already installed.

An Exit through a Basement Wall

1 **Drilling the hole.** Locate the hole by measuring from a reference point that is accessible from both sides of the wall. The hole must lie at least 3 inches from joists, sill plate and flooring of the house to provide clearance for a junction box.

Set an LB fitting against the wall where you intend to drill to be sure that the fitting does not overlap a siding joint. Verify the measurements you have made with a ¼-inch test hole, then bore through the joist, using a ⅞-inch spade bit.

In a cinder-block wall, make a hole with a ⅞-inch star drill. Hammer it through the center of a block in the second course below the siding, where blocks are hollow. Rotate the drill ⅛ turn after each hammer tap to make a neat, round hole.

2 **Installing the fitting.** Remove a knockout from the back of a junction box and mount the box over the hole you have drilled. Select a nipple long enough to reach through the wall and into the junction box, and screw it to an LB fitting. Insert the nipple into the hole, and bend conduit to run from the fitting into the trench (*Step 2, opposite*). Withdraw the fitting from the wall, screw it onto the conduit and push the nipple back through the wall. Outside the house, strap the conduit to the foundation and caulk around the nipple. Inside, secure the nipple with a star nut, then screw on a plastic bushing.

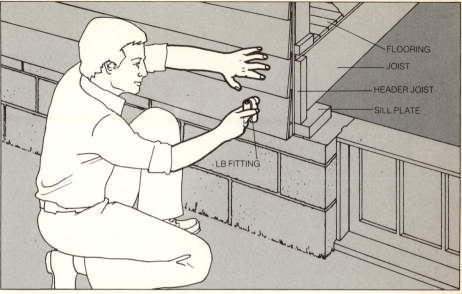

FLOORING
JOIST
HEADER JOIST
SILL PLATE
LB FITTING

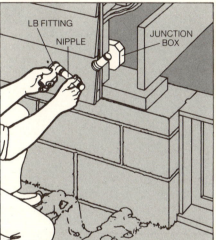

LB FITTING
NIPPLE
JUNCTION BOX

An Exit through an Eave

Installing a box on the overhang. Fasten together an outdoor box, a nipple, a corner elbow and a section of conduit. Hold the assembly against the soffit board, with the conduit against the house siding and the box between two rows of soffit nails. (You can make the conduit less conspicuous by running it alongside a downspout.) Use the box as a template to mark the soffit for a cable hole and for mounting-screw holes. Drill the holes with a ³/₃₂-inch bit for ¾-inch, No. 8 screws and a 1⅛-inch spade bit for the cable hole.

Fish a cable from an indoor circuit through the cable hole, attach a two-part connector (*page 85*) and screw the box to the connector. Mount the box on the soffit (because of the connector's shape, you may have to enlarge the cable hole with a rasp to align the mounting tabs with the screw holes). Strap the nipple to the soffit board and the conduit to the wall, and bend a section of conduit to run into the trench (*Step 2, opposite*).

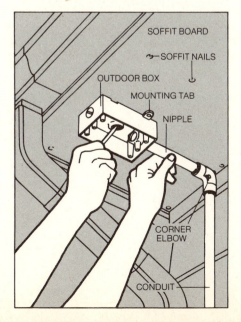

SOFFIT BOARD
SOFFIT NAILS
OUTDOOR BOX
MOUNTING TAB
NIPPLE
CORNER ELBOW
CONDUIT

Digging a Trench

A trench is excavated in two distinct stages. First, lay out the route with stakes and string to keep it straight between bends. Cut out the turf to a depth of about 1½ inches and set it on plastic sheets to keep the grass from taking root. While the sod is out on the ground, keep it barely damp.

In the second stage of the job, dig out the dirt and pile it to one side. A trench for UF cable, plastic conduit and thin-wall conduit should be about 8 inches wide and 1 foot deep; a trench for rigid conduit, 6 inches deep and about 4 wide. For long trenches, especially the deep ones for UF cable, consider renting a gasoline-powered trencher from a tool-rental agency.

If you must go under a sidewalk, dig to one side of the walk, then continue digging on the other side. Later, hammer one end of a length of conduit to a sharp, closed point, and drive the conduit beneath the walk to join the trenches, using a short-handled hammer, called a maul, that has a three-pound head. Then cut off the closed end of the conduit with a hacksaw so that you can join it to the rest of the conduit.

Completing an Outdoor System: The Fixtures

Most outdoor fixtures are simple to set up and they are all wired as they would be indoors. Some garden lights, for example, stick into the ground on a spike and plug into a receptacle. An outdoor light fixture can be hung from an eave; install a box on the overhang *(page 119)* and screw a light fixture to it.

A tall lamppost needs to be sunk a couple of feet into the ground. Many installers anchor the post in dirt reinforced with layers of stones so that it can be reset if it is accidentally knocked over. Most posts have a hole near the bottom for UF cable to enter, but if you install conduit, you must cut a slot in the pole wide enough to slip over the conduit.

Recessing a light in the brick wall of a stairway *(opposite)* requires the removal of a brick and the use of an unusual conduit fitting.

Electric-eye switches come in several designs and wattage capacities. Because such switches themselves operate on a tiny current, their wiring is slightly different from that of an ordinary switch. A freestanding box, often set up to house an outlying receptacle, is assembled in a specific sequence and must be provided with a cinder block and gravel base to protect it from accidental knocks.

When all the conduit, boxes and lampposts are in place, use a fish tape *(pages 10-11)* to pull UF cable or TW wires through the conduit. To ease passage, coat the wires with a wax-based lubricant, then fish all wires through at once. (Avoid petroleum-based lubricants; they can damage insulation.) Attach the wires to light fixtures, receptacles and switches as shown in Chapter 2, then use the continuity test on page 62 to check the

wiring before sealing the fixtures against the weather. Turn off current to the indoor circuit that will supply power, and connect the outdoor extension. Restore power and use the voltage tests on pages 21-23 for a final check. Fill in the trench —and the job is done.

Heating cables for the roof and strings of outdoor lights can be added after the wiring is finished. Heating cables come in kits, complete with clips to hold the cable in the zigzag pattern shown overleaf. The cable melts a path for water to run freely into the gutter and downspout, preventing ice build-up.

You can assemble strings of lights yourself *(overleaf)* using screw-on light fixtures called carnival sockets. These lights are not weatherproof; they should be taken down and stored as soon after the party as possible.

Setting Up a Lamppost

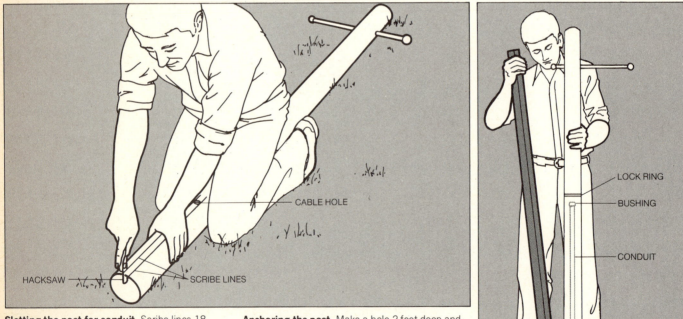

Slotting the post for conduit. Scribe lines 18 inches long and ⅞ inch apart on the post; they will usually extend beyond the UF cable hole in the post. Place the post on the grass and brace it between your knees. Cut along the scribed lines with a special hacksaw that grips the blade at one end and at the center (the frame of a regular hacksaw would get in the way). Bend back the ⅞-inch strip of metal between the cuts and saw it off. File away any sharp corners and edges. If the lamppost is in the middle of a run, cut a second slot on the opposite side of it so that the conduit can continue to the next fixture.

Anchoring the post. Make a hole 2 feet deep and about 8 inches in diameter. Unless you are installing UF cable, bend conduit so that it rises out of the ground at the center of the hole. To simplify fishing wire later, the conduit should reach almost to the lock ring of an adjustable post or almost to the top of a fixed-length post. Cap the conduit with a plastic bushing. Slide the post over the conduit and, with a 2-by-4, tamp alternating layers of dirt and stones around the post. Check the post's vertical alignment frequently, and fill only to the bottom of the trench; fill the rest of the hole later, along with the trench.

Recessing a Step Light

Securing the fixture. Extract a brick from the wall by chipping out the surrounding mortar and cracking the brick with a cold chisel, then bore a ⅞-inch hole through the inner wall with a star drill. Into the threaded opening in the back of the fixture, screw a nipple that is long enough to reach into the trench behind the wall.

If the fixture opening is for ¾-inch conduit, reduce the size of the hole to take a ½-inch nipple by threading a star nut and a reducing bushing —a ring threaded inside for a ½-inch conduit and outside to fit a ¾-inch hole—onto one end of the nipple. Tighten the two fittings together. Screw the nipple into the box and then push the fixture into the recess you have made for it.

Pack mortar around the nipple where it protrudes from the back of the wall. After the mortar dries, screw an outdoor box to the nipple. Thread the fixture leads through the nipple and into the box, where they will be connected to the wiring. Install a light bulb and mount the fixture cover plate.

Switches that Sense the Dark

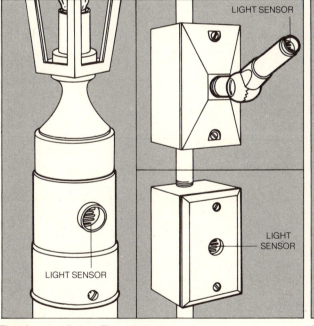

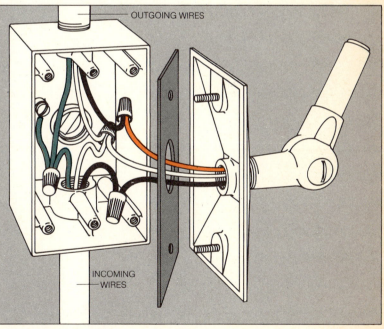

Electric-eye switches. These devices come in a variety of designs for different applications. In one version the light sensor that controls the switch is housed in a sleeve that fits over a lamppost. Another type is shaped like a jointed wand that can be pointed toward the sky; it screws into the cover plate used with an outdoor light fixture (*page 113*). A third type fits directly onto an outdoor box. These switches have a built-in time delay so they will not be affected by flashes from automobile headlights, and are made with different capacities for controlling bulbs up to 1,000 watts.

Connecting the switch. The three wires of an electric-eye switch are connected to circuit wires as illustrated above. Using wire caps, fasten the white switch wire to all the white wires in the box. Join the black switch wire to the incoming black wire and the red switch wire to the outgoing black one that supplies power to the light. Attach a green jumper to the box and the green incoming and outgoing wires. The switch will control any number of lights on the same outgoing circuit up to the capacity of the switch.

An Outlying Receptacle

Attaching an end-of-the-run box. Enlarge the trench to accommodate a cinder block 4 inches high. Bend a piece of conduit so that the end rises 8 inches above ground level, and screw on a box or fasten it with a threadless connector.

Attaching a middle-of-the-run box. Using a box with two conduit openings in one end, attach conduit as shown in the drawing at left. Then use a threadless connector to fasten a second leg of conduit to the other hole in the end of the box.

Making a sturdy base. Lower the cinder block over the box, then fill the space around the conduit legs with gravel to help prevent the box from wobbling. Install a GFI receptacle after you have fished wiring through the conduit.

Iceproofing the Eaves

Installing the cables. Use two heating-cable kits for each eave, one for the gutter and downspout, and one for the roof. The first cable should equal the combined lengths of gutter and downspout. To figure roof-cable length, multiply the gutter's length by 1.8, 2.6 or 3.5, depending on whether the cable is to go 1, 2 or 3 feet back from the edge.

Slip the cable clips under flexible asphalt shingles *(inset)*—or glue them to wood or slate shingles with epoxy cement —to position the roof cable to loop down into the gutter. Lay the other cable in the gutter, dropping one end into the downspout. Do not shorten the cables or let them touch or cross each other. Plug into an outdoor receptacle installed below the overhang, leaving a loop of wire so water will not run into the receptacle. Use an indoor switch and a ground-fault interrupter *(page 116)*. This shock protection is valuable—a short circuit could energize downspouts—although it is sometimes omitted to avoid false alarms caused by wind-driven snow.

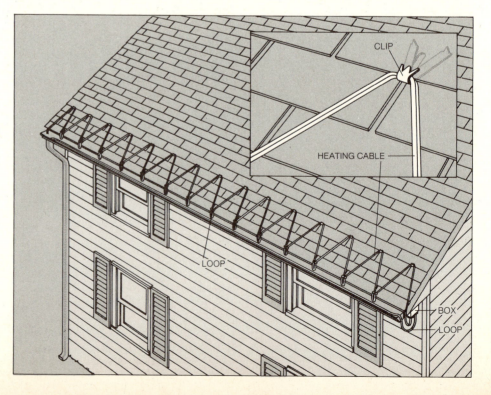

Carnival Lights for a Party

Stringing the wires. Party lights like the ones shown here consist of 12-gauge TW wires *(page 24)* that are hung from hooks or tied to trees and then fitted with clamp-on carnival light fixtures *(far right)* and plugged into an outdoor receptacle. In addition to the plug, sockets, and black and white wires, you need split insulators *(right)*, galvanized screw hooks and stout twine.

Caution: Carnival lights are not weatherproof and are for temporary use only. All parts except the plug must be placed out of reach and the wires should dip no lower than 7 feet above the ground.

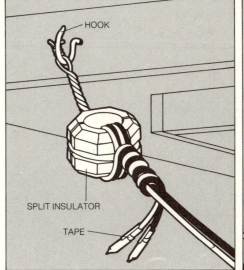

Supporting the wires. Hang wires on a hook by twisting TW wire around the halves of a split insulator, leaving a loop. Pass the conductor wires twice through the insulator; twist them around themselves. At a tree, tie the insulator on with twine, and thread wires through. Support the wires every 40 feet. Tape the far ends; they will carry current.

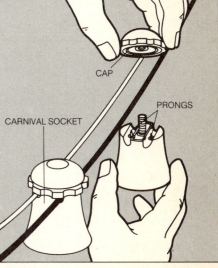

Attaching the sockets. Unscrew the cap from a carnival socket and, standing on a stepladder, position the socket so the brass prong lies opposite the black wire and the silver prong lies opposite the white wire. Screw the cap onto the wires so that the prongs will penetrate the insulation.

Picture Credits

The sources for the illustrations in this book are shown below.

Cover—Ken Kay. 6—Ken Kay. 10,11—Al Freni. 12,13—Drawings by Nicholas Fasciano. 14 through 17—Drawings by Whitman Studio, Inc. 18,19—Drawings by Vantage Art, Inc. 20—Drawings by John Sagan. 21,22,23—Drawings by Whitman Studio, Inc. 24,25—Drawings by Vantage Art, Inc. 26,27—Drawings by Peter McGinn. 28—Ken Kay. 30,31—Drawings by Vantage Art, Inc. 32,33—Drawings by Nicholas Fasciano. 34,35—Drawings by Raymond Skibinski. 37,38—Drawings by Nicholas Fasciano. 40,41—Drawings by John Sagan. 43 through 47—Drawings by Raymond Skibinski. 49—Drawings by Whitman Studio, Inc. 50,51—Drawings by Nicholas Fasciano. 52 through 59—Drawings by Vantage Art, Inc. 60—Ken Kay. 63—Drawing by Raymond Skibinski. 65 through 69—Drawings by Nicholas Fasciano. 70 through 75—Drawings by Adolph E. Brotman. 76,77—Drawings by Raymond Skibinski. 78 through 83—Drawings by Adolph E. Brotman. 84,85—Drawings by Raymond Skibinski. 86 through 89—Drawings by Nicholas Fasciano. 90 through 93—Drawings by Raymond Skibinski. 94 through 97—Drawings by Peter McGinn. 98 through 103—Drawings by Whitman Studio, Inc. 104,105—Drawings by Raymond Skibinski. 106,107—Drawings by Whitman Studio, Inc. 108—Ken Kay. 110,111—Drawings by Vantage Art, Inc. 112 through 115—Drawings by John Sagan. 116 through 123—Drawings by Whitman Studio, Inc.

Acknowledgments

The index/glossary for this book was prepared by Mel Ingber. The editors also thank the following: K. L. Bellamy, Chief Electrical Inspector, Ontario Hydro, Toronto, Ontario, Canada; W. S. Cahill, Supervisor, Press Relations, Lamp Business Division, General Electric Company, Cleveland, Ohio; Robert Hausler, McPhilben Lighting Co., Melville, N.Y.; Flynn E. Hudson, Manager, Product Engineering, Wiring Device Dept., General Electric Company, Providence, R.I.; Jan Kroeze, New York City; E. J. Papik, Manager of Technical Services, Canadian Electrical Manufacturers Association, Toronto, Ontario, Canada; John Poliak, Chief Engineer, Construction Material Division, Leviton Manufacturing Co., Inc., Little Neck, N.Y.; Albert Reed, State Superintendent, New York Board of Fire Underwriters, New York City; Phil Seltzer, New York City; Carole Stupell, Ltd., New York City; Wilford I. Summers, Secretary, National Electrical Code Committee, National Fire Protection Association, Boston, Mass.; The Amagansett Plant Shop, Amagansett, N.Y.

Index/Glossary